Optical Nanomanipulation

Optical Nanomanipulation

David L Andrews
University of East Anglia, Norwich, UK

David S Bradshaw
University of East Anglia, Norwich, UK

Morgan & Claypool Publishers

Rights & Permissions
To obtain permission to re-use copyrighted material from Morgan & Claypool Publishers, please contact info@morganclaypool.com.

ISBN   978-1-6817-4465-0 (ebook)
ISBN   978-1-6817-4464-3 (print)
ISBN   978-1-6817-4467-4 (mobi)

DOI   10.1088/978-1-6817-4465-0

Version: 20161201

IOP Concise Physics
ISSN 2053-2571 (online)
ISSN 2054-7307 (print)

A Morgan & Claypool publication as part of IOP Concise Physics
Published by Morgan & Claypool Publishers, 40 Oak Drive, San Rafael, CA, 94903 USA

IOP Publishing, Temple Circus, Temple Way, Bristol BS1 6HG, UK

To my wonderful daughter Elspeth and her partner Jack—David Andrews
To my grandparents, parents, sister and brother—David Bradshaw

Contents

Preface

Among the numerous phenomena that were discovered within the first decade of the laser, one of the most striking was the pioneering demonstration by Arthur Ashkin, working at Bell Laboratories, that a laser beam could controllably manipulate microparticles. It was a proof of principle that paved the way for the technological breakthrough that became known as optical tweezers, which opened up a whole range of research applications; it also led to the development of atom cooling and subsequent trapping techniques. In fact, the discovery proved to be a tipping point for the identification of a remarkably extensive variety of phenomena in which laser light can be utilised to manipulate micro- and nano-sized particles. The whole field has grown and developed almost beyond recognition over the last fifty years, and it now represents a cutting edge of modern optics.

Today, outside of the routine application of optical tweezers as a technique for cell sorting procedures in medical laboratories, there is an increasing trend towards using light to locate, manoeuvre, rotate and assemble a wide range of nanoparticles, microparticles and even large molecules. Meanwhile, at the research end of the spectrum, the availability of a much wider range of laser power and beam structures now facilitates a variety of other methods for handling individual atoms and small molecules. The non-contact nature of optical manipulation offers huge advantages for the investigation and fabrication of matter at nanoscale dimensions, and it enables many techniques that would be simply impossible using any other methodology.

This book provides a broad introductory survey of this remarkable field, aiming to establish and clearly differentiate its physical principles, and also to provide a snapshot portrait of many of the most prominent current applications. Primary emphasis is placed on developing an understanding of the fundamental photonic origin behind the mechanism that operates in each type of effect. To this end, the first few chapters introduce and develop core theory, focusing on the physical significance and source of the most salient parameters, and revealing the detailed interplay between the key material and optical properties. Where appropriate, both classical and photonic (quantum mechanical) representations are discussed. The number of equations is purposely kept to a minimum, and only a broad background in optical physics is assumed.

With copious examples and illustrations, each of the subsequent chapters then sets out to explain and exhibit the main features and uses of the various distinct types of mechanism that can be involved in optical nanomanipulation, including some of the very latest developments. To complete the scene, we also briefly discuss applications to larger, biological particles. Overall, this book aims to deliver to the non-specialist an amenable introduction to the technically more advanced literature on individual manipulation methods. Full references to the original research papers are given throughout, and an up-to-date bibliography is provided for each chapter, which directs the reader to other selected, more specialised sources. We freely acknowledge that the studies we describe represent only a fraction of the field as a whole; the subject has quickly outgrown the scale that any review could conceivably capture in its entirety. The responsibility for any errors is, of course, our own.

David L Andrews
David S Bradshaw
Norwich, October 2016

Acknowledgments

It is a great pleasure to acknowledge the privilege of help, in this endeavour, from many of the top researchers in this topic. We are especially indebted to numerous individuals who have supplied us with original figures, most of which have not been published previously. In addition we gratefully thank colleagues, especially Mathew Williams, for numerous helpful comments. We are greatly indebted to all.

Author biography

David L Andrews

David L Andrews is a Professor at the University of East Anglia in the UK and leads research on fundamental molecular photonics, energy harvesting and transport, optomechanical forces, quantum and nonlinear optics. He has over 350 research papers and fifteen books to his name. The current focus of his research group is on quantum aspects of optical transmission, optical vortices and chirality, frequency conversion, optical nanomanipulation and switching, optically nonlinear mechanisms in fluorescence, and energy transfer processes. Professor Andrews is a Fellow of the SPIE, the Optical Society of America, the Royal Society of Chemistry, and the Institute of Physics. He is also a member of the Board of Directors of SPIE.

David S Bradshaw

David S Bradshaw is an honorary research associate at the University of East Anglia in the UK. He graduated twice from the same university, first receiving a Master's degree in chemical physics (which included a year at the University of Western Ontario, London, Canada) and then a PhD in theoretical chemical physics. Overall, David has co-written 80 research papers, including book chapters, all based on molecular quantum electrodynamics. He has also created a website explaining the key physics in this theory. His long running interests include optical trapping, resonance energy transfer, optical binding and nonlinear optics. David is a Member of the Institute of Physics and the Royal Society of Chemistry.

List of symbols

$\langle\rangle_T$	time average
$\langle\rangle_R$	rotational average
$\mathbf{A}$	classical electromagnetic vector potential
$\hat{\mathbf{A}}$	electromagnetic vector potential operator
a	particle diameter
$\hat{a}$	photon annihilation operator
$\hat{a}^\dagger$	photon creation operator
$\mathbf{B}$	magnetic field vector
$\hat{\mathbf{B}}$	magnetic field operator
$\mathbf{b}$	magnetic polarisation unit vector
c	speed of light
c_d	drag coefficient
d	particle diffusion distance
$\mathbf{E}$	electric field vector
$\hat{\mathbf{E}}$	electric field operator
E_0	energy of initial particle state
E_I	energy of initial system state
E_r	energy of virtual particle state
$\mathbf{e}$	electric polarisation unit vector
$|F\rangle$	final system state
$f_{l,p}$	normalised radial function
g	acceleration due to gravity
H_{rad}	classical radiation Hamiltonian
$\hat{H}$	Hamiltonian operator of total system
$\hat{H}_0$	Hamiltonian operator of unperturbed system
$\hat{H}_{\mathrm{int}}$	interaction Hamiltonian operator
$\hat{H}_{\mathrm{mat}}$	matter Hamiltonian operator
$\hat{H}_{\mathrm{rad}}$	radiation Hamiltonian operator
h	Planck constant
$\hbar$	reduced Planck constant
$|I\rangle$	initial system state
I	irradiance (intensity) of light beam
I_ω	irradiance per unit frequency
$\mathbf{J}_{\mathrm{rad}}$	classical total angular momentum of radiation
$\hat{\mathbf{J}}_{\mathrm{rad}}$	total angular momentum operator of radiation
$\mathbf{k}$	wave-vector
k	magnitude of the wave-vector
k_B	Boltzmann constant
k_x	force constant for trap stiffness
L_{Abra}	orbital angular momentum of light through a medium (Abraham)
L_{Mink}	orbital angular momentum of light through a medium (Minkowski)
$\mathbf{L}_{\mathrm{rad}}$	classical orbital angular momentum of radiation
$\hat{\mathbf{L}}_{\mathrm{rad}}$	orbital angular momentum operator of radiation
L	left handed (circularly polarised light)
l	topological charge of vortex beam
ℓ	Fourier component index
M_{FI}	matrix element

m	particle mass
$\hat{N}$	number operator
N'	probability density for trapped particle position
n_a	atom number density
n_ω	refractive index
n_ϕ	number of absorbed photons
$\mathbf{P}_{\mathrm{rad}}$	classical linear momentum of radiation
$\hat{\mathbf{P}}_{\mathrm{rad}}$	linear momentum operator of radiation
P	radiation pressure
p	radial index
p_a	mean atomic momentum
R	reflectivity of the substance
R	right handed (circularly polarised light)
$|R\rangle$	intermediate system state
$\mathbf{r}$	position vector
r	off-axis radial distance
$|r\rangle$	virtual state of particle
r_b	radius of beam ring
$\mathbf{S}_{\mathrm{rad}}$	classical spin angular momentum of radiation
$\hat{\mathbf{S}}_{\mathrm{rad}}$	spin angular momentum operator of radiation
$\mathbf{s}$	inter-particle displacement
s	separation distance between particles
$|s\rangle$	excited state of two-level particle
T	absolute temperature
T_{c}	critical temperature
$T_{\min}$	minimum temperature
$\hat{T}_0$	resolvent operator
t	time
$\mathrm{tr}(\ )$	trace of a matrix
$\mathbf{u_k}$	unit vector of $\mathbf{k}$
$\mathbf{u_{kz}}$	unit vector of $\mathbf{k_z}$
$|u\rangle$	particle excited state
V	arbitrary volume
v	velocity
W	beam power
w_0	focused beam waist
z	axial position
α	scalar polarisability
α'	volume polarisability
α_ℓ	normal mode expansion variable
α_{ij}	transition polarisability tensor
Γ	process rate
γ	natural linewidth
ΔE	potential energy
∇	gradient operator (differential with respect to $\mathbf{r}$)
δv	recoil velocity
ε_0	permittivity of free space
η	viscosity of medium
η	polarisation state
λ	wavelength of trapping light

$\hat{\boldsymbol{\mu}}$	electric dipole moment operator
$\boldsymbol{\mu}^{0u}$	transition dipole moment (between ground and excited state)
ρ	density of particle
ρ_0	density of medium
ρ_F	density of states
σ_x	variance of a Gaussian distribution curve
$\boldsymbol{\tau}$	optical torque vector
Φ	electromagnetic scalar potential
ϕ	azimuthal angle
Ω_r	half-width at half-maximum of a Lorentzian lineshape
ω	angular optical frequency

Chapter 1

Nanomanipulation: why optical methods are best

Non-contact forces

In everyday experience, movement in the world we perceive is most often a result of contact forces. The slightest puff of breeze stirs the treetops only through direct physical contact between the leaves and the moving substance of the air. Such is the familiarity of the forces that we most commonly observe and deploy, that any occurrences of objects being physically moved without material contact seem intrinsically more exotic, and often rather enigmatic. Gravity, whose effects provide the most obvious example of producing motion without contact, was long considered a mystery; it succumbed to Newton's analysis without yielding much mechanistic insight, and it is indeed still frustratingly difficult to weave into a satisfactory Grand Unified Theory. At least for the present, gravity also remains a force beyond our capacity to control.

Over many centuries—at least since the first use of a lodestone—forces of electrical and magnetic origin have held a particular fascination. They too manifest a clearly non-contact nature, yet of a kind that is indeed amenable to control and application. The eventual dawning of an understanding that light is itself electromagnetic in nature thus provided a logical basis for expecting it to have a capacity to exert a mechanical force. Indeed, such an expectation is consistent with what we have more recently discovered—that at the atomic level, even what we would normally call 'contact' forces are also fundamentally electrodynamic in origin, manifesting in bulk behaviour the effect of hidden interactions between the elementary charges that comprise every piece of matter.

To discover demonstrable manifestations of light-induced forces nonetheless proved much more difficult than for electric or magnetic effects. Maxwell and Bartoli [1, 2], amongst others, showed how and why light should have a capacity to convey and confer momentum, but the experimental confirmation awaited studies by Lebedev [3] and by Nichols and Hull [4] at the turn of the twentieth century. These works confirmed the presence of such effects, but also verified that conventional sources would produce radiation forces of so little magnitude as to be scarcely amenable to measurement—let alone of meaningful practical use—and often completely masked by thermal forces. Indeed, it was only with the arrival of intense light sources in the form of lasers, in the early 1960s, that the determination and exercising of control over optical force first became a practicable proposition.

The rapid pace of development of laser systems quickly provided experimental access to previously unattainable levels of intensity, substantially as a result of the high degree of angular confinement and controllable direction in their optical emission. Exploiting this feature, Ashkin's experimental studies first revealed the remarkable possibilities that laser beams could controllably manipulate small, neutral particles of matter [5]. In this inaugural work, he overcame any problems with thermal forces by using relatively transparent particles (i.e. latex spheres) suspended in a relatively transparent medium (water). These spheres were observed to be simultaneously drawn to the beam axis, and accelerated in the direction of the light: optical manipulation was demonstrated for the first time. In addition, a narrow beam-width also proved significant in another sense; it provided for sharp intensity gradients—and hence differential forces—across the beam. This force is the basis for the renowned optical tweezer technique [6].

Issues of scale

The minimum physical width of an optical beam is generally prescribed by the Abbé diffraction limit [7], which is of the order of a wavelength—and therefore usually somewhere in the nanometre (sub-micron) regime for optical frequencies. (A variety of means have indeed been found to overcome this limit in super-resolution microscopy [8–11], notably the Nobel Prize winning technique of stimulated emission depletion, STED, developed by Hell et al [12, 13]). Consequently, the use of focused laser beams offers scope to exert controllable forces on micro- or nanometre-scale particles, including biological cells. Non-contact forces are especially useful in the nano regime because, on this scale, physical contact with manipulative equipment will often introduce problems due to adhesion, surface friction, stiction and contamination etc.

As we consider particles of progressively smaller size, gravity—whose effect essentially scales down with the cube of particle diameter—soon ceases to represent a significant obstacle. To completely offset its effect, micron sized particles can usually be stably suspended in a liquid—and this is indeed a common configuration for optical manipulation experiments. Here, viscous drag is the only major limitation to particle motion. (Intense light can in fact detach particles that are in physical contact, but the intensities required are much higher. Laser ablation operates on this principle [14]. However, such effects are not usually termed 'optical manipulation' and they lie outside the scope of this book.)

There are other problems to confront as the dimensional scale is further reduced to atomic and molecule-sized particles. Here, Brownian motion comes into play, serving

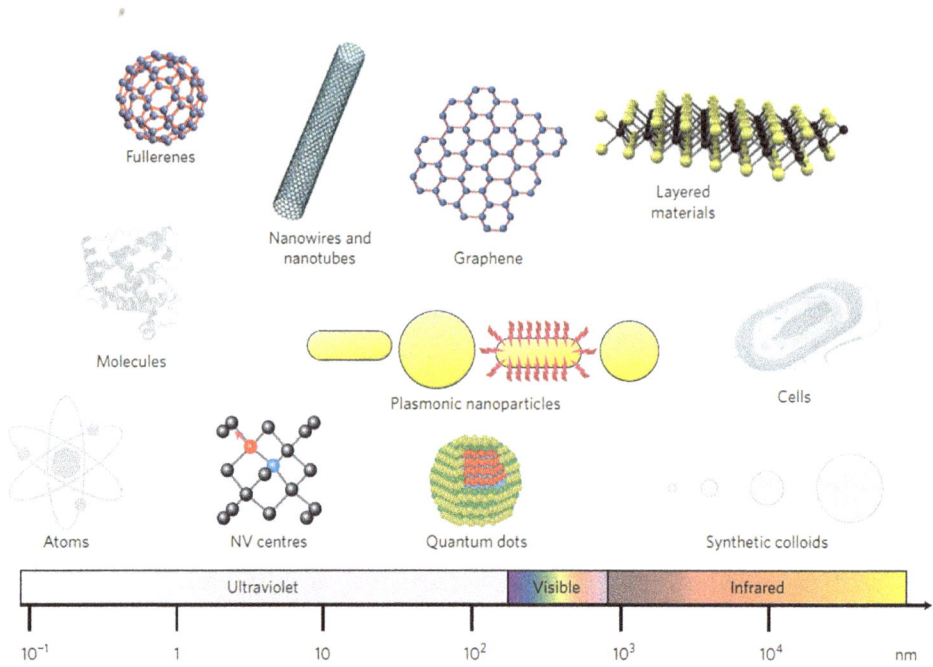

Figure 1.1. Optically trapped particles typically have three size regimes, exemplified by: atoms (a few Ångströms to a few nanometres), nanoparticles (a few nanometres to a few hundred nanometres) and microparticles (a fraction of a micron and above). The horizontal scale applies to both the size of the featured objects and the wavelength of the incident light; NV denotes nitrogen vacancy. Reproduced by permission from reference [15].

as a reminder that at nanoscale levels all matter is in continual flux. Even at a temperature approaching absolute zero there is a residual energy of motion, reflecting the intrinsic quantum mechanical zero-point energy of all matter. The thermal energy that registers temperature is associated with the random kinetic motions of atoms and molecules in the air, and other gases at ambient temperatures. For this reason, the optical manipulation of atoms and very small molecules is usually studied at the extremely low temperatures of ultra-cold gases (rather than liquid or solid states, since deposition or condensation would frustrate the sought particle mobility).

Due to the difficulties in scaling up ultra-cold techniques, or scaling down microparticle trapping procedures, particles of intermediate scale (including quantum dots, nanowires and nanotubes) are optically manipulated using different approaches. So, as we shall see, both the methodologies and types of optical force used for nanomanipulation exist in numerous different forms. The size range of particles that can be optically trapped is shown in figure 1.1.

References

[1] Bartoli A 1884 Il calorico raggiante e il secondo principio di termodinamica *Il Nuovo Cimento* **15** 193–202

[2] Maxwell J C 1954 *A Treatise on Electricity and Magnetism* vol 2 (New York: Dover Publications)

[3] Lebedev P N 1901 Experimental examination of light pressure *Ann. Phys.* (*Berlin*) **6** 433

[4] Nichols E F and Hull G F 1901 A preliminary communication on the pressure of heat and light radiation *Phys. Rev.* **13** 307–20

[5] Ashkin A 1970 Acceleration and trapping of particles by radiation pressure *Phys. Rev. Lett.* **24** 156–9

[6] Ashkin A, Dziedzic J M, Bjorkholm J E and Chu S 1986 Observation of a single-beam gradient force optical trap for dielectric particles *Opt. Lett.* **11** 288–90

[7] Abbe A 1873 Beiträge zur Theorie des Mikroskops und der mikroskopischen Wahrnehmung *Arch. f. Mikro. Anat.* **9** 413–8

[8] Hell S W and Stelzer E H K 1992 Properties of a 4Pi confocal fluorescence microscope *J. Opt. Soc. Am.* A **9** 2159–6

[9] Dunn R C 1999 Near-field scanning optical microscopy *Chem. Rev.* **99** 2891–928

[10] Gustafsson M G L 2000 Surpassing the lateral resolution limit by a factor of two using structured illumination microscopy *J. Microsc.* **198** 82–7

[11] Reymann J, Baddeley D, Gunkel M, Lemmer P, Stadter W, Jegou T, Rippe K, Cremer C and Birk U 2008 High-precision structural analysis of subnuclear complexes in fixed and live cells via spatially modulated illumination (SMI) microscopy *Chrom. Res.* **16** 367–82

[12] Hell S W and Wichmann J 1994 Breaking the diffraction resolution limit by stimulated emission: stimulated-emission-depletion fluorescence microscopy *Opt. Lett.* **19** 780–2

[13] Klar T A and Hell S W 1999 Subdiffraction resolution in far-field fluorescence microscopy *Opt. Lett.* **24** 954–6

[14] Shirk M D and Molian P A 1998 A review of ultrashort pulsed laser ablation of materials *J. Laser Appl.* **10** 18–28

[15] Maragò O M, Jones P H, Gucciardi P G, Volpe G and Ferrari A C 2013 Optical trapping and manipulation of nanostructures *Nat. Nanotechnol.* **8** 807–19

Chapter 2

Key properties of the radiation

Energy, linear momentum and angular momentum of light

Traditionally the analysis of optical forces begins with a text-book derivation based on the electromagnetic *Poynting vector* [1] (depicted in figure 2.1), working from its usual interpretation as the linear momentum density of a beam, to relate the local electromagnetic fields to the effects exerted on matter in the course of beam obstruction, deflection etc. Notwithstanding recent deliberations over the precise status of this interpretation—an issue which is still open to discussion—we too shall establish this link later in this chapter. Nevertheless in this book, as befits a focus on nanoscale interactions, it is appropriate to build upon a quantum foundation, i.e. a deeper and more fundamentally complete picture of light, before addressing the mechanical interactions of electromagnetic fields with matter. To this end, we start by considering the properties of photons and establish the factors that are relevant for producing an optical force.

Each photon can be considered to convey—in the sense of quantifiable attributes ascribed to matter acting as an absorber, for example—essentially three mechanical quantities: energy, linear momentum and angular momentum. As a scalar, of course, the radiation energy lacks any intrinsic directionality. However, for a beam delivering a particular photon flux, the linear and angular momenta conferred in the process of photon absorption afford the primary basis for radiation pressure, as well as linear forces and angular forces (torques). Other kinds of force, generally the secondary effects due to scattering, can be produced where the local potential energy landscape is shaped by a non-uniform beam intensity distribution.

A definition of each of these photon quantities, and the relationships between them, is now appropriate. Moreover, it proves helpful to introduce the classical beam representations alongside their quantum counterparts. The two are, of course, formally linked through a promotion of the classical expressions into operator form, from which the corresponding properties of individual photons emerge. For simplicity at this stage, the key equations are developed for free-space propagation;

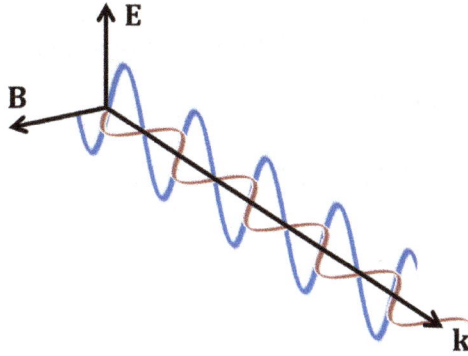

Figure 2.1. Plane polarised electromagnetic radiation: the **E**, **B** and **k** orthogonal axes denote the directions of the electric field, magnetic field and propagation direction, respectively. The Poynting vector (which is proportional to **E** × **B**) points in the **k** direction, the blue and red curves represent the electric and magnetic fields, respectively.

issues associated with local fields in the condensed phase are briefly addressed later. At this juncture, readers who have less interest in the following, concise summary of the formal development may skip ahead to chapter 3.

We begin with classical expressions for the generalised position- and time-dependent electric and magnetic free fields, $\mathbf{E}(\mathbf{r}, t)$ and $\mathbf{B}(\mathbf{r}, t)$ respectively, associated with electromagnetic radiation. Developing a Fourier series representation, each field can be written as a sum of two mode-summed contributions;[2]

$$\mathbf{E}(\mathbf{r}, t) = \mathbf{E}^{(+)}(\mathbf{r}, t) + \mathbf{E}^{(-)}(\mathbf{r}, t); \quad \mathbf{B}(\mathbf{r}, t) = \mathbf{B}^{(+)}(\mathbf{r}, t) + \mathbf{B}^{(-)}(\mathbf{r}, t), \quad (2.1)$$

whose constituent terms are referred to as the positive and negative frequency parts of the fields, respectively. The signs in the superscripts relate to spatial phase factors that feature in the explicit Fourier expansions: for example, the term $\mathbf{E}^{(+)}(\mathbf{r}, t)$ (also known as the *analytic signal*) is given by;

$$\mathbf{E}^{(+)}(\mathbf{r}, t) = \mathrm{i} \sum_{\ell} \mathbf{e}_\ell \mathcal{E}_\ell \alpha_\ell(t) \mathrm{e}^{\mathrm{i}(\mathbf{k}_\ell \cdot \mathbf{r})}, \quad (2.2)$$

where the Fourier component with index ℓ relates to a set of four numbers relating to the wave-vector $\mathbf{k}_\ell$ and polarisation unit vector $\mathbf{e}_\ell$, $\alpha_\ell(t)$ is a normal variable dependent on the positive form of the phase factor and $\mathcal{E}_\ell$ is a mode constant; $\mathbf{E}^{(-)}(\mathbf{r}, t)$ is the complex conjugate of the expression given by equation (2.2). Expressions for the two components of $\mathbf{B}(\mathbf{r}, t)$ have a similar form. To satisfy Maxwell's equations, the time-dependent factor $\alpha_\ell(t)$ has to carry a temporal phase factor $\exp(-\mathrm{i}\omega t)$. Thus, individual components of the Fourier series can be regarded as representing a beam of angular optical frequency $\omega = ck$, propagating with a wave-vector $\mathbf{k}$. For single-mode applications, the direction of $\mathbf{k}$ is usually taken to define the z-axis.

To determine analogous expressions in a quantum framework, i.e. where the electric and magnetic fields are quantised, the factor α_ℓ and its conjugate are promoted to the status of operators on radiation states—the operator form to be signified here, and henceforth, by the introduction of a carat (^) symbol—to obtain;

$$\hat{\mathbf{E}}(\mathbf{r}) = \hat{\mathbf{E}}^{(+)}(\mathbf{r}) + \hat{\mathbf{E}}^{(-)}(\mathbf{r})$$

$$= i \sum_{\mathbf{k},\eta} \left(\frac{\hbar c k}{2\varepsilon_0 V} \right)^{\frac{1}{2}} \left\{ \mathbf{e}^{(\eta)}(\mathbf{k}) \hat{a}^{(\eta)}(\mathbf{k}) e^{i(\mathbf{k} \cdot \mathbf{r})} - \overline{\mathbf{e}}^{(\eta)}(\mathbf{k}) \hat{a}^{\dagger(\eta)}(\mathbf{k}) e^{-i(\mathbf{k} \cdot \mathbf{r})} \right\}, \qquad (2.3)$$

$$\hat{\mathbf{B}}(\mathbf{r}) = \hat{\mathbf{B}}^{(+)}(\mathbf{r}) + \hat{\mathbf{B}}^{(-)}(\mathbf{r})$$

$$= i \sum_{\mathbf{k},\eta} \left(\frac{\hbar k}{2\varepsilon_0 c V} \right)^{\frac{1}{2}} \left\{ \mathbf{b}^{(\eta)}(\mathbf{k}) \hat{a}^{(\eta)}(\mathbf{k}) e^{i(\mathbf{k} \cdot \mathbf{r})} - \overline{\mathbf{b}}^{(\eta)}(\mathbf{k}) \hat{a}^{\dagger(\eta)}(\mathbf{k}) e^{-i(\mathbf{k} \cdot \mathbf{r})} \right\}, \qquad (2.4)$$

where the summation is taken over each photon with mode $(\mathbf{k}, \eta)$, in which η labels the polarisation state, c is the speed of light in a vacuum, ε_0 is the permittivity of free space, $\hbar$ is the reduced Planck constant and V is an arbitrary quantisation volume—a notional region within which the system of interest resides. The Hermitian conjugate pair of operators $\hat{a}^{(\eta)}(\mathbf{k})$ and $\hat{a}^{\dagger(\eta)}(\mathbf{k})$ signify photon annihilation and creation, respectively, i.e. they reduce or increase by one the number of photons present; the electric and magnetic polarisation vectors of each photon, denoted by $\mathbf{e}^{(\eta)}(\mathbf{k})$ and $\mathbf{b}^{(\eta)}(\mathbf{k}) = \mathbf{e}^{(\eta)}(\mathbf{k}) \times \mathbf{k}$ (overbars signifying their complex conjugates), also appear in the above expressions. At this juncture, it is noteworthy that the sum over polarisations can be effected over any pair that constitutes an orthogonal basis: this will usually be two orthogonal linear polarisations (often designated 'vertical' and 'horizontal' with respect to a reference plane) or left- and right-handed circular polarisation[1]. In the quantum formalism, $\hat{\mathbf{E}}(\mathbf{r})$ and $\hat{\mathbf{B}}(\mathbf{r})$ are not dependent on time since, in the assumed Schrödinger representation, any time dependence is held within the state vector.

The quantum energy operator is developed, in the usual way, from an expression for the classical Hamiltonian of the radiation, H_{rad}. The latter entails an integral of the energy density over volume, namely;

$$H_{\text{rad}} = \frac{\varepsilon_0}{2} \int_V \left(|\mathbf{E}(\mathbf{r}, t)|^2 + c^2 |\mathbf{B}(\mathbf{r}, t)|^2 \right) d^3 r. \qquad (2.5)$$

On substitution of the quantum field operators, i.e. equations (2.3) and (2.4), into this expression – where V again acquires the significance of a quantisation volume – the following Hamiltonian operator may be found;

$$\hat{H}_{\text{rad}} = \sum_{\mathbf{k},\eta} \left(\hat{N}^{(\eta)}(\mathbf{k}) + \frac{1}{2} \right) \hbar \omega, \qquad (2.6)$$

where the dependence of ω on $\mathbf{k}$ is left implicit; $\hat{N}^{(\eta)}(\mathbf{k}) = \hat{a}^{\dagger(\eta)}(\mathbf{k}) \hat{a}^{(\eta)}(\mathbf{k})$ is the *number operator*, which returns the number of photons when applied to a number (Fock)

---

[1] In fact, any two polarisation states that correspond to diametrically opposite positions on a Poincaré sphere may form an acceptable basis.

state. Moreover, for the ground state of the radiation, $\frac{1}{2}\hbar\omega$ represents the zero point energy per mode, associated with vacuum fluctuations that are present even in the absence of radiation. The latter is a feature that is a well-known characteristic of the quantum formalism. By inspection of equation (2.6), it can be determined that the energy of a single photon is $\hbar\omega$.

The classical linear momentum of the radiation, $\mathbf{P}_{rad}$, involves an integral of the Poynting vector, again taken over the volume V:

$$\mathbf{P}_{rad} = \varepsilon_0 \int_V (\mathbf{E}(\mathbf{r}, t) \times \mathbf{B}(\mathbf{r}, t))d^3\mathbf{r}, \qquad (2.7)$$

while the total angular momentum of the radiation, $\mathbf{J}_{rad}$, is proportional to an integral of the cross product between the Poynting vector and displacement, i.e.;

$$\mathbf{J}_{rad} = \varepsilon_0 \int_V \mathbf{r} \times (\mathbf{E}(\mathbf{r}, t) \times \mathbf{B}(\mathbf{r}, t))d^3\mathbf{r}. \qquad (2.8)$$

The above expression can, in turn, be re-written as the sum of two terms $\mathbf{J}_{rad} = \mathbf{S}_{rad} + \mathbf{L}_{rad}$, given explicitly as the respective terms in the following;

$$\mathbf{J}_{rad} = \varepsilon_0 \left(\int_V \mathbf{E}(\mathbf{r}, t) \times \mathbf{A}(\mathbf{r}, t) + E_i(\mathbf{r}, t)(\mathbf{r} \times \nabla)A_i(\mathbf{r}, t)d^3\mathbf{r} \right). \qquad (2.9)$$

Here, $\mathbf{A}(\mathbf{r}, t)$ is the electromagnetic vector potential, and the initial and final terms of the equation represent the spin and orbital angular momentum, respectively[2]. In the latter term, the repeated index i represents summation (also known as an *Einstein summation* [3]) over any frame of Cartesian coordinates.

The quantum operator form of the total angular momentum involves the following mode expansion of the vector potential:

$$\hat{\mathbf{A}}(r) = \sum_{\mathbf{k},\eta} \left(\frac{\hbar}{2\varepsilon_0 ck V} \right)^{\frac{1}{2}} \left\{ \mathbf{e}^{(\eta)}(\mathbf{k})\hat{a}^{(\eta)}(\mathbf{k})e^{i(\mathbf{k}\cdot\mathbf{r})} - \bar{\mathbf{e}}^{(\eta)}(\mathbf{k})\hat{a}^{\dagger(\eta)}(\mathbf{k})e^{-i(\mathbf{k}\cdot\mathbf{r})} \right\}. \qquad (2.10)$$

This potential is related to the electric and magnetic fields via $\mathbf{E} = -\partial\mathbf{A}/\partial t - \nabla\Phi$ and $\mathbf{B} = \nabla \times \mathbf{A}$, respectively, where ∇ denotes a differential with respect to $\mathbf{r}$ and Φ is the electromagnetic scalar potential: equations of precisely the same form apply to the corresponding quantum operators[3]. Inserting the field operators (2.3), (2.4) and (2.10) into the classical expressions $\mathbf{P}_{rad}$, $\mathbf{S}_{rad}$ and $\mathbf{L}_{rad}$ (as required) results in the

---

[2] In principle, the form of separation given by equation (2.9) is not unique, since the electromagnetic vector potential is not gauge invariant. In the Coulomb gauge used here, the separation is most effective on the assumption that optical beams have paraxial form.

[3] The operator of equation (2.10) is written in the Schrödinger representation, and is thus time-independent. As a result, it is inappropriate to use the first (time derivative) term in the electric field expression with equation (2.10) directly; the expression $i\hbar\dot{\hat{\mathbf{A}}} = [\hat{\mathbf{A}}, \hat{H}]$ delivers the necessary derivative.

production of their quantum equivalents. Namely, after a little manipulation, we find that;

$$\hat{\mathbf{P}}_{rad} = \sum_{\mathbf{k},\eta} \hat{N}^{(\eta)}(\mathbf{k})\hbar\mathbf{k}, \tag{2.11}$$

in which the zero-point momentum for each mode, $\frac{1}{2}\hbar\mathbf{k}$, vanishes due to each $\mathbf{k}$ in the summation having a matching $-\mathbf{k}$ (c.f. the zero-point energy, which persists since energy is always a positive quantity). Accordingly, it is established that the linear momentum is $\hbar\mathbf{k}$ for any photon with wave-vector $\mathbf{k}$. It also emerges that radiation states of pure circular polarisation are eigenstates of the operator for quantum spin, $\mathbf{S}_{rad}$. This is consistent with an angular momentum of $\pm\hbar$ per photon with the sign determined by handedness; the vector quantity directed along $\mathbf{k}$ is thus expressible as $\pm\hbar\mathbf{u}_{\mathbf{k}}$, where $\mathbf{u}_{\mathbf{k}}$ represents a unit vector of $\mathbf{k}$. This spin feature is concisely expressed by the following identity;

$$\hat{\mathbf{S}}_{rad} = \sum_{\mathbf{k}} \left\{ \hat{N}^{(L)}(\mathbf{k}) - \hat{N}^{(R)}(\mathbf{k}) \right\} \hbar\mathbf{u}_{\mathbf{k}}, \tag{2.12}$$

with the summation specifically cast in terms of circular polarisations, either left- or right-handed light as denoted by the superscript L and R. The latter expression shows, as might be expected, that the spin-angular momentum operator, $\hat{\mathbf{S}}_{rad}$, registers the disparity of left- and right-handed photon populations [4, 5]. The explicit result for $\hat{\mathbf{L}}_{rad}$ is deferred until chapter 8, where the appropriate mode structures are introduced.

In summary, the quantum development of theory shows that in addition to an energy of $\hbar\omega$, each photon conveys a linear momentum $\hbar\mathbf{k}$ and, in the case of circular polarisations, a unit spin angular momentum of $+\hbar$ for left-handed and $-\hbar$ for right-handed light.

Light inside a medium

The previous section implicitly assumed that the light traverses a vacuum. To account for electromagnetic radiation propagating within a medium or condensed phase, it is expedient to develop the fields (in either a classical or quantum form) to include a Lorenz factor, $(n_\omega^2 + 2)/3$ [6, 7][4], where n_ω is the refractive index of the dielectric medium. The momentum of light in a dielectric material is nonetheless the source of a century-old dispute known as the Abraham–Minkowski controversy. The debate centres on the Minkowski prediction that an increase in optical linear momentum should arise on entry of an electromagnetic wave into a dielectric medium, while Abraham predicted a decrease: the underlying logic is encapsulated in figure 2.2. Experiment has proven to be incapable of distinguishing between the two possibilites [8, 9], and for a long time theoretical studies were unable to agree on the correct solution [10, 11].

---

[4] In the quantum case, these factors are formally derived from first principles.

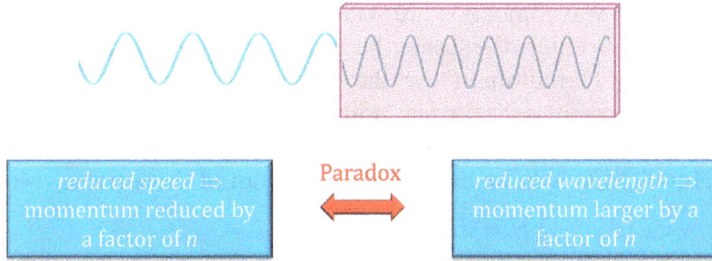

Figure 2.2. Illustration depicting the paradoxically opposite interpretations by Abraham (left-hand box) and Minkowski (right-hand box) of the momentum change in light entering a medium.

The controversy is now considered partially resolved. Since the momentum of a photon in a dielectric medium is an abstract concept that is experimentally unmeasurable, only the total momentum of the material and light combined is detectable—therefore, both predictions may be viewed as applicable. As a result, the division of total momentum into component (optical and material) parts is arbitrary and may be performed to produce a kinetic or canonical momentum. In fact, the Abraham and Minkowski formulations directly relate to the corresponding kinetic and canonical momenta [12–14]. Despite the recent progress, confusion connecting theory to experiment remains [15].

The two corresponding formulations for orbital angular momentum are $L_{\text{Mink}} = l\hbar$ and $L_{\text{Abra}} = l\hbar/n_\omega^2$, where l is the topological charge (detailed in chapter 8) and the former equation is identical to the free space form. The spin angular momentum of circularly polarised light will, in an isotropic medium, differ in the same way in their dependence on n_ω. The correct formalism for angular momentum is seemingly still debatable: Padgett *et al* support the Abraham formulation [16], although they acknowledge that an earlier paper favours the Minkowski form, while Pfeifer *et al* claim that both predictions are, in practice, identical [10].

Matter and its interaction with light

So far, only the character of the electromagnetic radiation has been examined. Before describing the optical forces that light may exert on materials, the key properties of matter and its interaction with light are now scrutinised. Alongside obvious mechanical features—for instance, mass and moment of inertia—there are electrodynamical properties, such as polarisability. The latter concerns the electric response of the delocalised electron cloud of the material, which has a quantum mechanical basis. Material properties of such a quantum origin are directly measurable; moreover, they are generally expressible by formulae that allow for their calculation, based on the fundamental principles of quantum mechanics. In anticipation of later detail, it is useful to recognise the typical treatment of such features.

Let us consider a system in which optical radiation engages with matter in some fashion. The usual starting point of a quantum analysis is to write down an expression for the Hamiltonian $\hat{H}$, i.e. the energy operator for the complete system. We recall that in simple quantum mechanics the 'system' usually comprises matter alone, and the state wavefunctions are secured as eigenstates of the corresponding $\hat{H}_{mat}$. When radiation enters the scene, the system Hamiltonian also contains a term $\hat{H}_{rad}$, which is the Hamiltonian for the radiation. However, whenever optical interactions occur there is a further additional term that needs consideration, namely the interaction Hamiltonian $\hat{H}_{int}$. Thus, the full quantum expression for the system Hamiltonian can be expressed as the sum of three parts;

$$\hat{H} = \hat{H}_{mat} + \hat{H}_{rad} + \hat{H}_{int}. \tag{2.13}$$

The last of these three terms is usually treated as a perturbation on the light and matter states. Theory, based on this perturbation, is then developed from the following general expression for the *matrix element*, M_{FI}, for progression from an initial system state $|I\rangle$ to a final state $|F\rangle$;

$$M_{FI} = \sum_{q=0}^{\infty} \langle F | \left\{ \hat{H}_{int} \left(\hat{T}_0 \hat{H}_{int} \right)^q \right\} | I \rangle. \tag{2.14}$$

In the above equation $\hat{T}_0$, known as a *resolvent operator*, is given by $\hat{T}_0 \approx (E_I - \hat{H}_0)^{-1}$, where the unperturbed system is represented by $\hat{H}_0 = \hat{H}_{mat} + \hat{H}_{rad}$ and E_I is

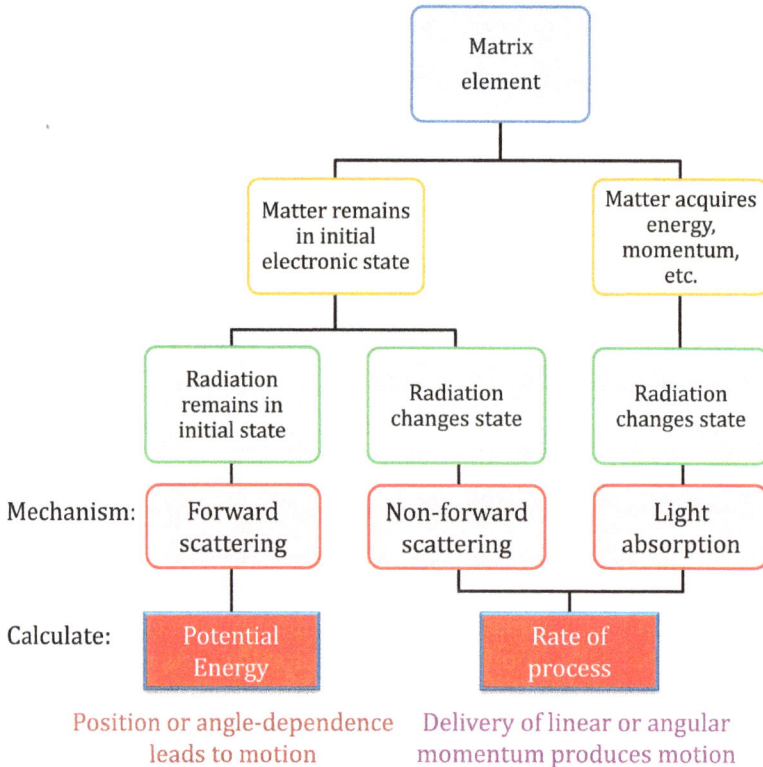

Figure 2.3. Flowchart exhibiting essential differences in implementation of the matrix element. The different pathways of interpretation are determined by considering any change in the quantum state of the matter and radiation; the relevant mechanism and its measurable are then decided.

the energy of the initial system state. A flowchart that assists the identification of the physical observable of a given matrix element is presented in figure 2.3.

Following application of equation (2.14), all of the key electrodynamical properties can be derived, and also analytical expressions for the observables associated with optical processes—such as rates, intensities and forces. For example, the $q = 0$ term delivers transition dipole moments and oscillator strengths for absorption processes, while the $q = 1$ term gives an explicit expression for polarisability, as we shall see below. In the latter connection, one clear value of these methods is the identification of the precise dependence on the frequency of the irradiating light. The importance of all these factors in the determination of optical forces will emerge in the following chapters.

References

[1] Poynting J H 1884 On the transfer of energy in the electromagnetic field *Philos. Trans. R. Soc.* A **175** 343–61

[2] Grynberg G, Aspect A and Fabre C 2010 *Introduction to Quantum Optics: From the Semi-Classical Approach to Quantized Light* (Cambridge: Cambridge University Press) p 314

[3] Einstein A 1916 Die grundlage der allgemeinen relativitätstheorie *Ann. Phys. (Berlin)* **354** 769–822

[4] Andrews D L and Coles M M 2012 Measures of chirality and angular momentum in the electromagnetic field *Opt. Lett.* **37** 3009–11

[5] Coles M M and Andrews D L 2012 Chirality and angular momentum in optical radiation *Phys. Rev.* A **85** 063810

[6] Juzeliūnas. G 1996 Microscopic theory of quantization of radiation in molecular dielectrics: normal-mode representation of operators for local and averaged (macroscopic) fields *Phys. Rev.* A **53** 3543–58

[7] Juzeliūnas. G 1997 Microscopic theory of quantization of radiation in molecular dielectrics. II. Analysis of microscopic field operators *Phys. Rev. A* **55** 929–34

[8] Padgett M J 2008 On diffraction within a dielectric medium as an example of the Minkowski formulation of optical momentum *Opt. Express* **16** 20864–68

[9] Zhang L, She W, Peng N and Leonhardt U 2015 Experimental evidence for Abraham pressure of light *New J. Phys.* **17** 053035

[10] Pfeifer R N C, Nieminen T A, Heckenberg N R and Rubinsztein-Dunlop H 2007 Momentum of an electromagnetic wave in dielectric media *Rev. Mod. Phys.* **79** 1197–216

[11] Barnett S M and Loudon R 2010 The enigma of optical momentum in a medium *Philos. Trans. R. Soc.* A **368** 927–39

[12] Barnett S M 2010 Resolution of the Abraham-Minkowski dilemma *Phys. Rev. Lett.* **104** 070401

[13] Baxter C and Loudon R 2010 Radiation pressure and the photon momentum in dielectrics *J. Mod. Opt.* **57** 830–42

[14] Milonni P W and Boyd R W 2010 Momentum of light in a dielectric medium *Adv. Opt. Photon* **2** 519–53

[15] Kemp B A 2011 Resolution of the Abraham-Minkowski debate: implications for the electromagnetic wave theory of light in matter *J. Appl. Phys.* **109** 111101

[16] Padgett M J, Barnett S M and Loudon R 2003 The angular momentum of light inside a dielectric *J. Mod. Opt.* **50** 1555–62

Chapter 3

Optically induced mechanical forces

The electric and magnetic fields of light may generate optical forces due to their engagement with the charge distributions in suitably irradiated nanoparticles. The effect is most obvious under resonance conditions, where a photon energy matches the quantum change in electronic energy of its absorber: the resulting optical transition will generally result in a modification of the internal energy distribution, at the atomic or molecular level. However, electromagnetic fields can also exert a more subtle effect under non-resonance conditions. Here, as we shall see, their capacity to produce short-lived *virtual* transitions lies at the heart of other important kinds of optical force. In these cases, light–matter interactions result in only transient changes to the internal electronic energy of the material. Then, according to the location of the matter relative to the local distribution of the light, forces may result from the position-dependent energy. In terms of photonics, optical trapping and manipulation of neutral particles by a laser beam may, therefore, involve two completely independent mechanisms.

Radiation forces

One form of optical force, the subject of this section, is traditionally associated with radiation pressure. As established in the previous chapter, electromagnetic radiation has an intrinsic linear momentum determined from the Poynting vector. When the light is absorbed, or the direction of the radiation is altered by its interaction with matter, light momentum is imparted onto the material in the form of a *radiation force*. Perhaps misleadingly, the term *scattering force* is sometimes applied to both forms of radiation force.

Classical description. For many years following the original discovery, no practical applications were considered feasible for the extremely small radiation forces associated with terrestrial-level solar or artificial light. Such effects were only considered necessary to account for the much larger solar radiation pressures observed in an astronomical context [1–6]. With the advent of continuous-wave laser beams, however, much larger magnitudes became available, to the extent that optical forces would soon be routinely used to accelerate, decelerate, deflect, guide and trap micro- and nano-particles [7]. It is useful to recognise the connection between the essentially classical, macroscopic and more obviously quantum level nanoscopic forces associated with optical radiation.

On the macroscopic scale, assuming the irradiation of a smooth, flat and opaque material, the original Maxwell–Bartoli force can readily be derived. Since an optical force can be described by both energy per unit distance and momentum per unit time (from Newton's second law) then it follows that optical momentum can be defined as energy divided by the speed of light. The radiation pressure, P, is the rate of change of momentum divided by the irradiated area so that: $P = I\cos^2\theta/c$, where I is the irradiance—the energy arriving at the surface per unit time (input power) per unit area, and θ is the angle between the surface normal and the incident radiation. This expression describes the radiation pressure acting on a perfectly absorbing surface, which is illustrated by figure 3.1(a). If the surface is perfectly reflecting, then a factor of two is included on the right-hand side of the expression; the latter is explained by the fact that a momentum is exerted by both the incident and reflected light, as shown by figure 3.1(b). In practice, the pressure delivered to a surface will lie somewhere between these two extremes, according to its reflectivity R. Hence, the general Maxwell–Bartoli expression emerges as follows:

$$P = (1 + R)(I/c)\cos^2\theta. \tag{3.1}$$

Quantum description. At the quantum level, the radiation force when n_ϕ photons are absorbed at a nanoparticle is found by multiplying the rate of absorption, Γ, by the linear momentum of the photons, $n_\phi\hbar\mathbf{k}$. The rate of absorption, for a given wave-vector $\mathbf{k}$ and polarisation η, is found from the Fermi rule, $\Gamma = (2\pi\rho_F/\hbar)|M_{FI}|^2$, where ρ_F is the density of states and M_{FI} is the matrix element determined from first-order time-dependent perturbation theory [8]. The result emerges from the $q = 0$ term in equation (2.14) as:

$$M_{FI} = \langle F|\hat{H}_{\text{int}}|I\rangle = \langle(n-1); u\,|-\hat{\boldsymbol{\mu}}\cdot\hat{\mathbf{E}}(\mathbf{r})|n; 0\rangle. \tag{3.2}$$

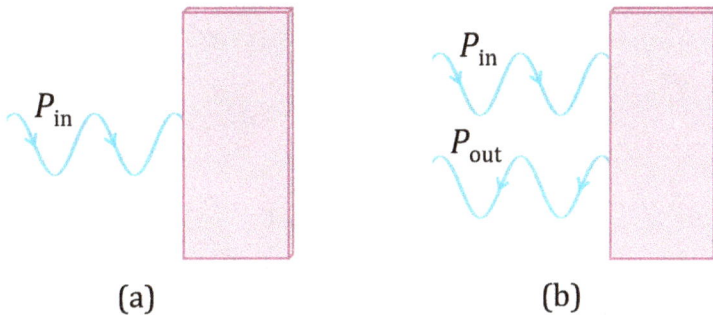

(a)   (b)

Figure 3.1. Radiation pressure, determined from the Maxwell–Bartoli expression, applied to a surface due to (a) absorption ($R = 0$) and (b) reflection ($R = 1$): P_{in} and P_{out} represent the radiation pressure due to the incident and reflected light, respectively.

Here, using Dirac bracket notation, $|I\rangle = |n; 0\rangle$ and $|F\rangle = |(n-1); u\rangle$ represent the specific initial and final system states, respectively, decomposed into radiation and particle components: the initial state contains n photons and the particle is in the ground state 0, and the final state (which follows photon absorption) comprises $n-1$ photons and the particle in its excited state u. At this juncture, it is appropriate to introduce the Feynman diagrammatic representation; the diagram corresponding to equation (3.2) is shown in figure 3.2.

The structure of equation (3.2) reveals that it is a change in state of the radiation field that allows for the transfer of momentum from the irradiating light to the nanoparticle. Also appearing in equation (3.2) is the explicit expression for the interaction Hamiltonian, i.e. $\hat{H}_{int} = -\hat{\mu} \cdot \hat{E}(r)$ where $\hat{\mu}$ is the electric dipole moment operator that acts on the particle states to produce a transition dipole moment μ^{u0} and $\hat{E}(r)$ acts on the radiation states. Here, in fact, only the $\hat{E}^{(+)}(r)$ portion of equation (2.3) is required since it is this term that includes the photon annihilation operator. Assuming $e \| \mu^{u0}$ and with the density of states given by $\rho_F = I_\omega V/2\pi cn\hbar^2\omega$ [9], the radiation force is determined as;

$$\mathbf{F} = \frac{n_\phi I_\omega \mathbf{k}}{2c\varepsilon_0 \hbar} \left| \mu^{u0} \right|^2, \tag{3.3}$$

where I_ω is the irradiance per unit frequency. In cases where the incident photons are deflected, rather than absorbed, the treatment of theory is a little more intricate and involves second-order perturbation theory.

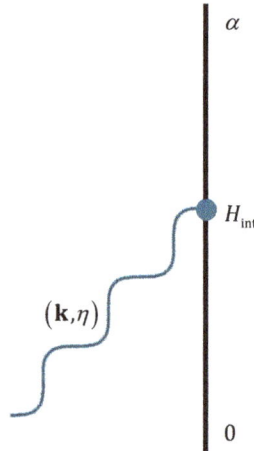

Figure 3.2. Feynman diagram illustrating the radiation force due to single photon absorption. The blue wavy line represents a photon of mode $(\mathbf{k}, \eta)$, the vertical line symbolises an optical centre that begins in a ground state 0 and finishes in its excited state α, the blue dot denotes the light–matter interaction, labelled by H_{int}, and time travels upwards. Since the initial state $|I\rangle = |n; 0\rangle$ and final state $|F\rangle = |(n-1); u\rangle$ differ, this diagram represents a process in which the physical observable is a rate, delivered from the matrix element using the Fermi rule.

The other elementary optical process that can, in principle, impart momentum at nanoscale dimensions is the time-inverse of photon absorption, i.e. emission. However, when emission is a spontaneous phenomenon in free space, the reactive force that acts on the emitter does not produce any overall effect due to the random directions in which such emission occurs.

Gradient force

Another form of optical mechanism that results in the centre-of-mass motion of a free particle is commonly known as the *gradient force* (or, alternatively, the *dipole force*). At the atomic or molecular level this involves the production of intensity-dependent internal energy level shifts, associated with the dynamic (ac) Stark effect. This arises from the local perturbation of electronic wavefunctions by the electric field of the radiation, through an interaction in which the latter features quadratically. In such a case, since the level shifts depend on the intensity of the light, an effective potential energy surface is introduced wherever that intensity varies with the particle position—which is usually across the cross-section of a beam and, thus, is related to the transverse field distribution. The gradient force is determined from the spatial derivative of the potential energy.

Classical description. In terms of classical electrodynamics, the gradient force is commonly defined in terms of the Lorentz force on a point dipole which, introducing a constant of proportionality (a scalar polarisability) α, is expressed as;

$$\mathbf{F}(\mathbf{r},\ t) = \alpha\left(\frac{1}{2}\nabla E(\mathbf{r},\ t)^2 + \frac{\mathrm{d}}{\mathrm{d}t}(\mathbf{E}(\mathbf{r},\ t) \times \mathbf{B}(\mathbf{r},\ t))\right). \tag{3.4}$$

The second term relates to the Poynting vector—which can again be cast, using equation (2.1), as a sum of four terms involving $\mathbf{E}^{(+)}(\mathbf{r},\ t)$ and $\mathbf{E}^{(-)}(\mathbf{r},\ t)$ in cross-products with their magnetic counterparts. It then transpires that the product contributions $\mathbf{E}^{(+)}(\mathbf{r},\ t)\mathbf{B}^{(-)}(\mathbf{r},\ t)$ and its complex conjugate are explicitly time-independent, and the two other terms carry factors $\exp(\pm 2i\omega t)$, hence oscillating too rapidly to engage with whole-particle motion. The 'Poynting' term in equation (3.4) thus disappears, and the time-averaged optical force (denoted below by $\langle\rangle_T$) becomes [10];

$$\langle\mathbf{F}(\mathbf{r})\rangle = \frac{\alpha}{2}\nabla\langle E(\mathbf{r})^2\rangle_{\mathrm{T}}. \tag{3.5}$$

Here, the electric field is involved in a quadratic response and, therefore, we can conclude that the gradient force is sustained in an oscillating field (static fields often relate to a linear response).

Quantum description. Turning again to the quantum theory, it is apparent that in the gradient force (unlike the radiation force) the electromagnetic field suffers no change in system state, i.e. no transfer of momentum arises. Here, the position-dependent potential energy, ΔE, is evaluated as an expectation value, in which the initial and final system states are identical, so that the interaction involves photon annihilation and recreation into the same radiation mode—this mechanism is known as forward-Rayleigh (elastic) scattering, as depicted in figure 3.3, wherein the frequency of the trapping beam is non-resonant with any energy level transitions in the nanoparticle.

Interpreting figure 3.3, using the framework of second-order perturbation theory i.e. the $q = 1$ term in equation (2.14), the following matrix element signifies the evaluation of an expectation value;

$$\Delta E = \text{Re } M_{II} = \text{Re}\left\{\sum_R \frac{\langle I|\hat{H}_{\text{int}}|R\rangle\langle R|\hat{H}_{\text{int}}|I\rangle}{E_I - E_R}\right\}, \qquad (3.6)$$

where $|I\rangle = |n; 0\rangle$ denotes both the initial and final system state and $|R\rangle = |n - 1; r\rangle$ is an intermediate system state: E is the energy of the state denoted by the subscript. It is important to note that $|R\rangle$ involves a short-lived virtual particle state r, rather than any specific excited state u. Moreover, $\hat{H}_{\text{int}} = -\hat{\boldsymbol{\mu}} \cdot \hat{\mathbf{E}}(\mathbf{r})$ will invoke either term of equation (2.3) so that either $\hat{\mathbf{E}}^{(+)}(\mathbf{r})$ or $\hat{\mathbf{E}}^{(-)}(\mathbf{r})$ appears in $\hat{H}_{\text{int}}$, depending on whether the operator acts on a photon annihilation or creation event. Using a beam irradiance (in units of W m^{-2}) given by $I = n\hbar c^2 k/V$, the gradient force is found as;

$$\mathbf{F}(\mathbf{r}) = \left(\frac{\nabla I(\mathbf{r})}{2\varepsilon_0 c}\right)\alpha_{ij}(\omega)\bar{e}_i e_j. \qquad (3.7)$$

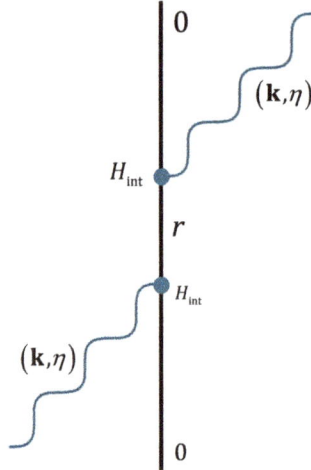

Figure 3.3. Feynman diagram illustrating the gradient force based on forward-Rayleigh scattering, in which a photon is annihilated and another recreated into the same radiation mode. The optical centre begins and ends in the ground state, traversing through a short-lived virtual state r. Since the initial and final state are identical, i.e. $|I\rangle = |F\rangle = |n; 0\rangle$, this diagram represents a potential energy obtained from the real part of the matrix element.

Here, the η and $\mathbf{k}$ dependences are suppressed on the unit polarisation vector $\mathbf{e}$, and the polarisability $\alpha_{ij}(\omega)$ emerges, in a complete form, as a second rank property tensor explicitly given by;

$$\alpha_{ij}(\omega) = \sum_r \left\{ \frac{\mu_i^{0r} \mu_j^{r0}}{E_{r0} - \hbar\omega} + \frac{\mu_j^{0r} \mu_i^{r0}}{E_{r0} + \hbar\omega} \right\}, \qquad (3.8)$$

where $E_{0r} = E_0 - E_r$. In passing we note that results are more often cast in terms of the linear optical susceptibility, in place of the polarisability tensor, when light is scattered by larger nanoparticles.

Equation (3.7) is written in a concise form that once again uses the Einstein summation convention (for repeated tensor indices). In any application to three-dimensional space, it simply means that where subscript indices i, j are used to signify x, y or z components in a Cartesian frame of reference, the repetition of an index within a given term signifies an implicit summation over those same three directions. For example, a scalar product $\boldsymbol{\mu} \cdot \mathbf{E}$ may be written as $\mu_i E_i \equiv \mu_x E_x + \mu_y E_y + \mu_z E_z$ and the scalar $\alpha_{ii} \equiv \alpha_{xx} + \alpha_{yy} + \alpha_{zz} = \mathrm{tr}(\boldsymbol{\alpha})$. (To save any confusion, the rule also stipulates that no such index can appear more than twice in a given term.) It transpires that equation (3.7) tallies with equation (3.5) through a rotational average leading to $\langle \alpha_{ij}(\omega) \bar{e}_i e_j \rangle_R = \frac{1}{3}\mathrm{tr}(\boldsymbol{\alpha}) = \alpha$ together with the relation $I = c\varepsilon_0 |\mathbf{E}|^2$[1].

The orientational effect of the gradient force is less well-known, and it is not manifested if the target particles have spherical symmetry—nor can it be identified from the equations if a scalar representation of the polarisability is used. The majority of the micro- and nano-particles studied in optical traps are indeed of essentially spherical form, meaning that their polarisability tensor can be represented by a diagonal matrix with all three components identical. Nonetheless, there is interest in non-spherical particles, and the simplest case arises with particles of cylindrical or ellipsoidal symmetry. Here, the diagonal polarisability matrix has one value, $\alpha_\parallel$, for the axial direction, and another, $\alpha_\perp$, for each of the other two axes[2]. Any such particle, initially sitting at an arbitrary angle with respect to the linearly polarised input radiation, will experience a torque that tends to align its long axis with the polarisation plane, at right-angles to the direction of light propagation. However, it is noteworthy that larger microscale particles with cylindrical symmetry will align with the propagation direction, in a beam of

---

[1] Molecular polarisability is often reported as a 'volume polarisability' defined as $\alpha' = \alpha/4\pi\varepsilon_0$ and, as such, the force per molecule that follows from equation (3.7) is of the order $2\alpha' I/cw_0$, where w_0 is the (focused) beam waist. For example $\alpha' = 10^{-29}$ m^3, $I = 10^{15}$ W m^{-2} and $w_0 = 1$ μm gives an optical force of around 0.1 fN, within measurable range using equipment such as atomic force microscopy.

[2] It is noteworthy that $\alpha_\parallel$ will always have the larger magnitude in a prolate ellipsoid since, in the longer axial direction, the polarizability—which is, basically, a measure of the shift in electron distribution to engender a dipole moment—will correspond to a larger electron shift.

narrow waist, to maximise engagement with the high intensity region [11]. An expression for the torque at the nanoscale is given by;

$$\boldsymbol{\tau} = (\alpha_{\perp} - \alpha_{\parallel})(E_y\mathbf{i} - E_x\mathbf{j})E_z, \qquad (3.9)$$

where E_x, E_y and E_z are the electric field directed in the x-, y- and z-directions: $\mathbf{i}$ and $\mathbf{j}$ respectively signify x- and y-unit vectors.

There is another less widely recognised mechanical effect of light, known as *electrostriction* [12], which occurs within the bulk of a solid through which light passes. It too depends on the electric field of the radiation engaging with the charges of which the material is comprised. As illustrated in figure 3.4, this effect can cause a small shrinkage in one direction and expansion in another, according to the orientation of its electric field. Its internal effect within nanoscale particles is too small to be significant, but the underlying mechanism is closely related to the process of *optical binding*, which relates to interactions *between* nanoparticles. The latter is a very important phenomenon in the field of optical nanomanipulation: we shall return to this topic in chapter 12.

To summarise; in terms of a quantum framework, the crucial difference in the calculation of the radiation and gradient force is that the former uses the rate as the physical observable, since the initial and final system states differ, while the latter is found from the potential energy in which the initial and final states are identical [13]—as indicated by the flowchart of figure 3.4. The radiation force involves the transfer of momentum from the beam to the nanoparticle through either light absorption or non-forward Rayleigh scattering (where the incident and emergent photon have the same energy, but travel in different directions). The gradient force implies an attraction of the nanoparticle to the high intensity part of the beam via forward Rayleigh scattering (all photons here have the same energy and direction).

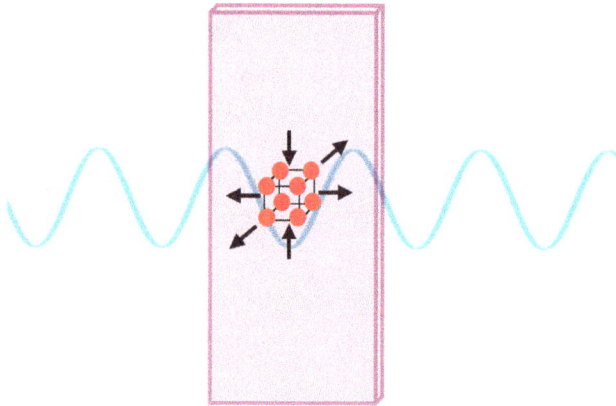

Figure 3.4. Electrostriction produced by light travelling through a solid. The constituent atoms of the solid are compressed together in the direction of the electric field (up and down arrows) and expanded in the other perpendicular directions (left and right arrows, in and out of plane arrows).

References

[1] Poynting J H 1904 Radiation in the solar system: its effect on temperature and its pressure on small bodies *Philos. Trans. R. Soc.* A **202** 525–52

[2] Robertson H P 1937 Dynamical effects of radiation in the solar system *Mon. Not. R. Astron. Soc.* **97** 423–37

[3] Ives N E 1963 The effect of solar radiation pressure on the attitude control of an artificial earth satellite, R. & M., No. 3332

[4] Tsander F A 1964 *Problems of Flight by Jet Propulsion: Interplanetary Flights* (Jerusalem: Israel Program for Scientific Translations Ltd)

[5] Burns J A, Lamy P L and Soter S 1979 Radiation forces on small particles in the solar system *Icarus* **40** 1–48

[6] McInnes C R 2004 *Solar Sailing: Technology, Dynamics and Mission Applications* (Berlin: Springer)

[7] Ashkin A 1997 Optical trapping and manipulation of neutral particles using lasers *Proc. Natl Acad. Sci. USA* **94** 4853–60

[8] Cohen-Tannoudji C, Dupont-Roc J and Grynberg G 1992 *Atom-Photon Interactions: Basic Processes and Applications* (New York: Wiley) p 59

[9] Craig D P and Thirunamachandran T 1998 *Molecular Quantum Electrodynamics: An Introduction to Radiation-Molecule Interactions* (New York: Dover) p 89

[10] Chaumet P C and Nieto-Vesperinas M 2000 Time-averaged total force on a dipolar sphere in an electromagnetic field *Opt. Lett.* **25** 1065–7

[11] Simpson S H and Hanna S 2007 Optical trapping of spheroidal particles in Gaussian beams *J. Opt. Soc. Am.* A **24** 430–43

[12] He G S 2015 *Nonlinear Optics and Photonics* (Oxford: University Press) p 95

[13] Bradshaw D S and Andrews D L 2013 Interparticle interactions: Energy potentials, energy transfer, and nanoscale mechanical motion in response to optical radiation *J. Phys. Chem.* A **117** 75–82

Chapter 4

Laser cooling and trapping of atoms

Doppler cooling

In 1933, Frisch first demonstrated that radiation pressure could act on atoms by deflecting an atomic sodium beam with light from a lamp [1]. Much later, in 1975, the possibility of such an effect leading to the cooling of atoms by laser radiation was independently proposed by Hänsch and Schawlow [2] and also by Wineland and Dehmelt [3], and other developments then quickly followed [4–10]. The initial experiments on the topic involved the cooling of a cloud of Mg^{2+} ions confined to a Penning trap [11] and the cooling of trapped Ba^+ ions [12].

Laser (or Doppler) cooling relates to the radiation force and its operation in atoms is described as follows. Electromagnetic radiation irradiates an atom with a frequency 'detuned to the red', i.e. the light frequency is slightly lower than a prominent transition within the atom—the latter usually assumed to be a two-level system. Resonance conditions are met, resulting in photon absorption, when the atom is moving towards the light source—since the radiation becomes blue-shifted (relative to the atom) due to the Doppler effect. The momentum of the atom is then lowered by an amount equal to the momentum of the absorbed photon, $\hbar\mathbf{k}$. The newly excited atom soon relaxes back into the unexcited state via spontaneous emission. The recoil radiation that results is emitted in a random direction and, hence, it is highly unlikely to be equal to $\hbar\mathbf{k}$. Applying many absorption-emission events to an ensemble of atoms that is irradiated by two counter-propagating beams (a 1D set-up), the momentum from the recoil radiation averages to zero and the net momentum transferred to the atoms is $n_\phi\hbar\mathbf{k}$ in a direction opposite to atomic motion, producing a damping force. Therefore, the velocities of the atoms are reduced and the temperature lowered. It is worth emphasising that the mechanism at work here is specifically a two-step process of photon absorption and emission; it is not scattering.

Doppler cooling limit. In the cycle of repeated absorption and spontaneous emission events, the mean reaction force resulting from the emissions is zero; hence the time-averaged recoil velocity $\langle \delta v \rangle = 0$. Indeed, the mean velocity of the trapped atom itself becomes zero, i.e. $\langle v \rangle = 0$. Nonetheless the quadratic mean $\langle |v|^2 \rangle$ need not vanish and, additionally, there are statistical fluctuations as the atom experiences the effects of individual photon absorption and emission events. Accordingly, since $\langle |v|^2 \rangle$ is non-zero, the atoms have a residual kinetic energy, as first observed by Bjorkholm *et al* [13]. This imposes a lower limit on the achievable extent of cooling, known as the *Doppler cooling limit*. An expression to determine the latter is written as $T_{min} = \hbar\gamma/2k_B$ where T_{min} is the minimum temperature, k_B is the Boltzmann constant and γ is the natural linewidth of the atomic transition [14]. For example, an atomic gas at a pressure of 1 Torr, with a natural linewidth of 10 MHz, would give a limit of 50 μK. This expression was experimentally verified by Chang *et al* using metastable helium-4 gases in the Doppler regime [15].

Optical molasses

Once the Doppler cooling limit is reached, diffusive motion of the ultracold atoms is observed [16–19]. There is a laser beam configuration—comprising, in three dimensions, three pairs of counter-propagating beams (figure 4.1)—that produces a loose assembly of diffusing ultracold atoms; this is known as *optical molasses*, so named because the beams are thought to produce something that resembles a 'viscous fluid'. Optical molasses, which was initially used to confine and cool sodium atoms, was first observed by Chu *et al* [16]. An important facet in the experiment is the provision of a frequency chirp [20], i.e. a rapid shift in light frequency with time. This frequency alteration counters the change in the Doppler shift as the atoms slow, enabling continuous resonance absorption of light by the atoms. In the late 80s, following the discovery that the Doppler cooling limit could be overcome [21, 22],

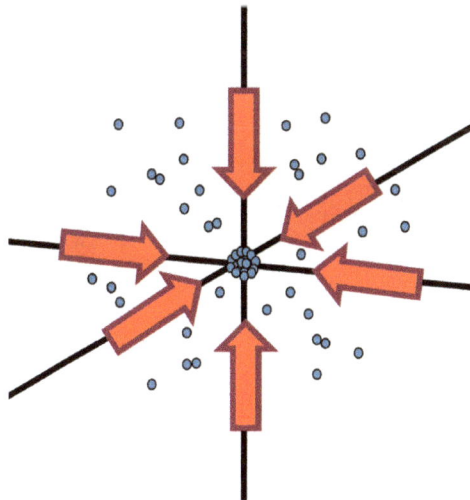

Figure 4.1. Schematic for the laser cooling of atoms within an optical molasses set-up. The red arrows denote pairs of counter-propagating beams and the circles signify atoms; a representation of a condensed cloud of atoms is centred at the intersection of the lasers.

it was recognised that other cooling mechanisms may occur in the system. This realisation led to the development of sub-Doppler laser cooling techniques such as Sisyphus cooling [23].

Bose–Einstein condensates. A combination of sub-Doppler laser cooling—the Nobel Prize-winning technique developed by Chu, Cohen-Tannoudji and Phillips—and evaporative cooling can produce a state of matter known as a Bose–Einstein condensate (BEC) [24–26], in which multiple ultracold atoms together enter a single, lowest energy quantum state. Such a state is only achievable for particles with a net spin that has an integer value (these are bosons). In 1995, the first BEC was produced when a vapour of rubidium-87 atoms condensed into a BEC at 170 nanokelvin [27], an achievement that led to the award of the Nobel Prize to Cornell and Wieman[1]. The formation of a BEC requires that the ultracold sample has a de Broglie wavelength (h/p_a, where p_a is the mean atomic momentum associated with thermal motion and h is Planck's constant) that is larger than the average spacing between atoms, at a temperature below a critical limit. For an ideal (non-interacting) Bose gas, this critical temperature T_c is determined from the expression $T_c = 3.31 \hbar^2 n_a^{2/3} / m k_B$ where m is the mass of each atom and n_a is the atom number density [28]. While a sub-Doppler temperature is achievable in optical molasses, evaporative cooling techniques [29–31] can further decrease the temperature so that the critical limit can be attained. Evaporative cooling works on the principle that, following confinement of the atoms in a magneto-optical trap, the lowering of the sides of the potential well (in which the trapped atoms reside) allows the most energetic atoms to escape. The result is that the remaining atoms contain lower kinetic energies and, thus, a lower temperature. Cooling by evaporation has the dual effect of increasing the density of the trapped atoms and lowering the temperature, both crucial in the production of BECs. The formation of a BEC is illustrated in figure 4.2.

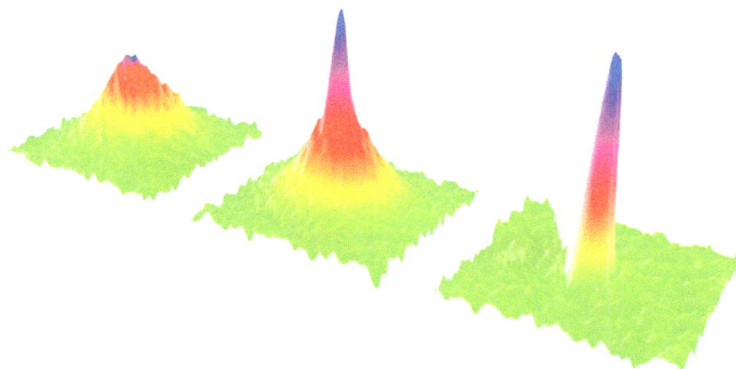

Figure 4.2. False-colour images displaying the density distribution for the cross-section of a cloud after the release of atoms from a trapping potential: (left) just above the phase transition, (centre) partly condensed, and (right) an almost pure condensate. Adapted from www.quantum-munich.de/media/Bose-Einstein-condensates. Reproduced with permission from Max Planck Institute of Quantum Optics (MPQ).

---

[1] Ketterle was a co-recipient of the Nobel Prize with Cornell and Wieman. Soon after the creation of the first condensate, he headed a group that condensed a gas of sodium atoms into a BEC.

Diatomic molecules. Doppler cooling requires the continuous absorption-emission cycle of a very large number of photons in a closed two-level transition. In practice, atomic systems rarely behave as simple two-level systems. For example, sodium atoms have two hyperfine ground sub-levels: for the purpose of driving transitions to achieve cooling, one is useful and the other is not. If spontaneous decay to unwanted states occurs, the two-level absorption-emission cycle will be compromised and ultracold temperatures may become unobtainable. In such a scenario, an effective two-level cycle is secured by re-pump lasers that excite the 'lost' population back into the cycle [32]. Unlike atoms, diatomic molecules contain vibrational and rotational states. As a result, application of the re-pump technique to these molecules is extremely difficult to implement, requiring an impractical number of re-pump lasers of various frequencies. Therefore, the laser cooling of diatomic molecules within optical molasses is essentially unviable. Although the production of ultracold molecules by laser cooling has been convincingly demonstrated using a configuration of lasers acting on strontium monofluoride [33, 34]—and, more recently, yttrium oxide [35] and calcium fluorine [36]—less than a dozen diatomic molecules have the internal structure amenable for similar laser cooling methods.

References

[1] Frisch O R 1933 Experimenteller Nachweis des Einstenschen Strahlungsruckstosses *Z. Phys.* **86** 42–8

[2] Hänsch T W and Schawlow A L 1975 Cooling of gases by laser radiation *Opt. Commun.* **13** 68–9

[3] Wineland D J and Dehmelt H 1975 Proposed $10^{14} \, \Delta\nu < \nu$ laser fluorescence spectroscopy on Tl$^+$ mono-ion oscillator *Bull. Am. Phys. Soc.* **20** 637

[4] Letokhov V S, Minogin V G and Pavlik B D 1977 Cooling and capture of atoms and molecules by a resonant light field *Sov. Phys. JETP* **45** 698–705

[5] Wineland D J and Itano W M 1979 Laser cooling of atoms *Phys. Rev.* A **20** 1521–40

[6] Javanainen J 1980 Light-pressure cooling of trapped ions in three dimensions *Appl. Phys.* **23** 175–82

[7] Javanainen J and Stenholm S 1980 Broad band resonant light pressure II: Cooling of gases *Appl. Phys.* **21** 163–7

[8] Letokhov V S and Minogin V G 1981 Laser radiation pressure on free atoms *Phys. Rep.* **73** 1–65

[9] Stenholm S 1986 The semiclassical theory of laser cooling *Rev. Mod. Phys.* **58** 699–739

[10] Phillips W D, Gould P L and Lett P D 1988 Cooling, stopping, and trapping atoms *Science* **239** 877–83

[11] Wineland D J, Drullinger R E and Walls F L 1978 Radiation-pressure cooling of bound resonant absorbers *Phys. Rev. Lett.* **40** 1639–42

[12] Neuhauser W, Hohenstatt M, Toschek P and Dehmelt H 1978 Optical-sideband cooling of visible atom cloud confined in parabolic well *Phys. Rev. Lett.* **41** 233–6

[13] Bjorkholm J E, Freeman R R, Ashkin A and Pearson D B 1980 Experimental observation of the influence of the quantum fluctuations of resonance-radiation pressure *Opt. Lett.* **5** 111–3

[14] Lett P D, Phillips W D, Rolston S L, Tanner C E, Watts R N and Westbrook C I 1989 Optical molasses *J. Opt. Soc. Am.* B **6** 2084–107

[15] Chang R, Hoendervanger A L, Bouton Q, Fang Y, Klafka T, Audo K, Aspect A, Westbrook C I and Clément D 2014 3D laser cooling at the Doppler limit *Phys. Rev. A* **90** 063407

[16] Chu S, Hollberg L, Bjorkholm J E, Cable A and Ashkin A 1985 3D viscous confinement and cooling of atoms by resonance radiation pressure *Phys. Rev. Lett.* **55** 48–51

[17] Phillips W D, Prodan J V and Metcalf H J 1985 Laser cooling and electromagnetic trapping of neutral atoms *J. Opt. Soc. Am.* B **2** 1751–67

[18] Sesko D, Fan C G and Wieman C E 1988 Production of a cold atomic vapor using diode-laser cooling *J. Opt. Soc. Am.* B **5** 1225–7

[19] Hodapp T W, Gerz C, Furtlehner C, Westbrook C I, Phillips W D and Dalibard J 1995 3D spatial diffusion in optical molasses *Appl. Phys.* B **60** 135–43

[20] Ertmer W, Blatt R, Hall J L and Zhu M 1985 Laser manipulation of atomic beam velocities: demonstration of stopped atoms and velocity reversal *Phys. Rev. Lett.* **54** 996–9

[21] Lett P D, Watts R N, Westbrook C I, Phillips W D, Gould P L and Metcalf H J 1988 Observation of atoms laser cooled below the Doppler limit *Phys. Rev. Lett.* **61** 169–72

[22] Shevy Y, Weiss D S, Ungar P J and Chu S 1989 Bimodal speed distributions in laser-cooled atoms *Phys. Rev. Lett.* **62** 1118–21

[23] Dalibard J and Cohen-Tannoudji C 1989 Laser cooling below the Doppler limit by polarization gradients: simple theoretical models *J. Opt. Soc. Am.* B **6** 2023–45

[24] Bose S N 1924 Plancks Gesetz und Lichtquantenhypothese *Z. Phys.* **26** 178–81

[25] Einstein A 1924 Quantentheorie des einatomigen idealen Gases *Sitzungsber. phys.-math. Kl* 261–67

[26] Einstein A 1925 Quantentheorie des einatomigen idealen Gases. Zweite Abhandlung *Sitzungsber. phys.-math. Kl.* 3–14

[27] Anderson M H, Ensher J R, Matthews M R, Wieman C E and Cornell E A 1995 Observation of Bose–Einstein condensation in a dilute atomic vapor *Science* **269** 198–201

[28] Bagnato V, Pritchard D E and Kleppner D 1987 Bose–Einstein condensation in an external potential *Phys. Rev. A* **35** 4354–8

[29] Hess H F 1986 Evaporative cooling of magnetically trapped and compressed spin-polarized hydrogen *Phys. Rev. B* **34** 3476–9

[30] Tommila T 1986 Cooling of spin-polarized hydrogen atoms trapped in magnetic-field minima *Europhys. Lett.* **2** 789–95

[31] Masuhara N, Doyle J M, Sandberg J C, Kleppner D, Greytak T J, Hess H F and Kochanski G P 1988 Evaporative cooling of spin-polarized atomic hydrogen *Phys. Rev. Lett.* **61** 935–8

[32] Phillips W D 1998 Laser cooling and trapping of neutral atoms *Rev. Mod. Phys.* **70** 721–41

[33] Shuman E S, Barry J F, Glenn D R and DeMille D 2009 Radiative force from optical cycling on a diatomic molecule *Phys. Rev. Lett.* **103** 223001

[34] Shuman E S, Barry J F and DeMille D 2010 Laser cooling of a diatomic molecule *Nature* **467** 820–3

[35] Hummon M T, Yeo M, Stuhl B K, Collopy A L, Xia Y and Ye J 2013 2D magneto-optical trapping of diatomic molecules *Phys. Rev. Lett.* **110** 143001

[36] Zhelyazkova V, Cournol A, Wall T E, Matsushima A, Hudson J J, Hinds E A, Tarbutt M R and Sauer B E 2014 Laser cooling and slowing of CaF molecules *Phys. Rev. A* **89** 053416

Chapter 5

Dielectric and metal nanoparticles: Rayleigh regime

Arthur Ashkin and optical tweezers

From 1978 onwards, Ashkin demonstrated that neutral dielectric microparticles (and atoms) can be trapped with single-beam techniques that utilise radiation forces [1–3]. Notwithstanding these advances, it became clear that to optically trap particles of any significantly smaller dimensions, i.e. nanoparticles, it would be essential to use the gradient force [4]. In 1986, the first single-beam trap based on the gradient force was produced by Ashkin *et al* [5]—a set-up now known as *optical tweezers*. Under such a scheme, dielectric particles of sizes ranging from 10 μm down to ~25 nm were shown to be stably trapped in water. The success of optically trapping particles with dimensions at the lower end of this scale represented the first achievement of optical nanomanipulation in the *Rayleigh-scattering regime*—a term signifying a particle diameter much smaller than the wavelength of the trapping light (i.e. $a \ll \lambda$), with scattering properties relating to the polarisability [6, 7].

Ashkin *et al* predicted that their new technique would 'open a new size regime to optical trapping encompassing macromolecules, colloids, small aerosols, and possibly biological particles'. In fact, a year later, this group was the first to use optical tweezers to trap and manipulate viruses, bacteria and single cells [8, 9]. Building on this foundation, it is now possible to trap red blood cells *in vivo* [10]. Ashkin's remark proved to be prophetic, since substances of these types are precisely those trapped by present-day optical tweezer techniques. It is interesting that the development of optical tweezers has been much more rapid than for the radiation pressure traps of previous years, stemming from the realisation by the science community that such a technique leads to a wide range of applications. By developing the optical tweezers, Ashkin (pictured in figure 5.1) once again proved to be the definitive pioneer of optical trapping and manipulation.

doi:10.1088/978-1-6817-4465-0ch5

Figure 5.1. The founding father of optical trapping and manipulation Arthur Ashkin (right) with colleague Joseph Dziedzic. They were the first to optically trap biological substances. Reproduced with permission from AIP Emilio Segre Visual Archives, Physics Today Collection.

Optical tweezers work on the principle that a highly focused continuous-wave laser beam [11][1], typically with a Gaussian (TEM$_{00}$ mode) intensity profile, can optically trap and manipulate micro- and nano-sized dielectric particles. In such techniques, created by using a high-quality microscope objective, the particles are trapped within the narrow beam waist of the focused beam. While the gradient force attracts the particles laterally towards the high intensity part of the beam at its centre, the radiation force acts to 'push' the particles in the direction of beam propagation. However, the possibility of axial displacement from the focal region is countered by a diminution of intensity, due to the beam divergence, away from this region. The result is that the particle remains in a fixed axial position along the beam. Modern devices (for example, the BioPhotonics Workstation [12] of figure 5.2) operate under the control of computers, which are able to measure displacements and forces with high precision and accuracy.

An alternative to the single-beam configurations is the dual counter-propagating beam set-up, analogous to the first optical trap constructed in 1970 by Ashkin [13], in which the radiation forces are (to a large extent) balanced out by the two oppositely facing beams. This form of optical trapping was revolutionised by a fibre optics technique created in 1993 by Constable *et al* [14], who demonstrated that such a scheme may provide a number of enhanced features, including the ability to trap large cells. A few years later Zemanek *et al* designed a counter-propagating geometry, wherein an incident beam and its reflection interfere to produce a Gaussian standing wave trap, that operates at powers considerably smaller than optical tweezers when trapping nanoparticles [15]. However, while these forms of

---

[1] Most optical tweezers use monochromatic continuous-wave laser sources that are minimally absorbed by the trapped material. However, the use of a pulsed, picosecond or femtosecond laser in an optical tweezers set-up is possible, based on a fast pulse repetition rate (typically in the GHz range).

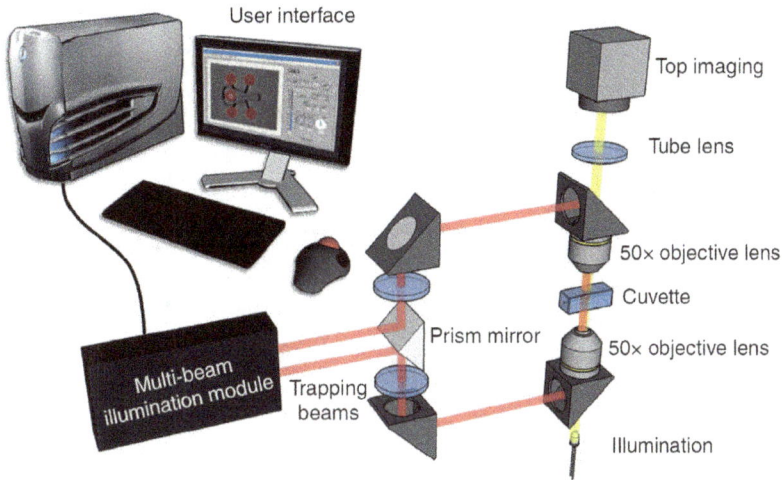

Figure 5.2. An example of a modern, computer-controlled optical trapping system. Known as the BioPhotonics Workstation, this set-up generates counter-propagating beam traps using a multi-beam illumination module. The top and bottom set of counter-propagating beams are imaged in the cuvette through opposing 50× objective lenses. The top imaging is fed real-time to the user-interface for intuitive optical manipulation. Reproduced by permission from reference [12].

trapping require only two weakly focused beams, experimental complexities are added compared to the single beam trap, such as constraints in terms of the precision optical alignment.

Optical trapping of nanoparticles

The optical trapping and manipulation of dielectric, semiconductor and metal nanoparticles are now discussed. In the three decades that followed the initial construction of the optical tweezers, experimental studies most commonly focused on the trapping of micro-sized particles such as beads and cells. However, in present times, the optical trapping of much smaller, nanoscale-dimension particles has become a rapidly growing field [16–19]. An effective deterrent to technical interest in trapping techniques for nanoparticles stems from the complexities in their implementation; the application of a sufficient optical force to a particle of nanoscale dimensions presents a challenge.

Dielectric and semiconductor nanoparticles. The trapping of nanoparticles is much more readily achievable when at least one of the particle dimensions extends into the micron size range. Examples include bundles of carbon nanotubes [20–23] (trapping of individual nanotubes remains a problematic task), semiconductor nanowires [24–30], nanorods [31, 32] and 2D graphene flakes [33, 34]; however, quantum dots have also been optically trapped [35, 36]. The process of angular orientation of an optically trapped nanorod is illustrated in figure 5.3. In modelling terms, while a nanoparticle (such as a small molecule) can be treated within the electric dipole approximation—i.e. the particle is represented as a point electric dipole in an

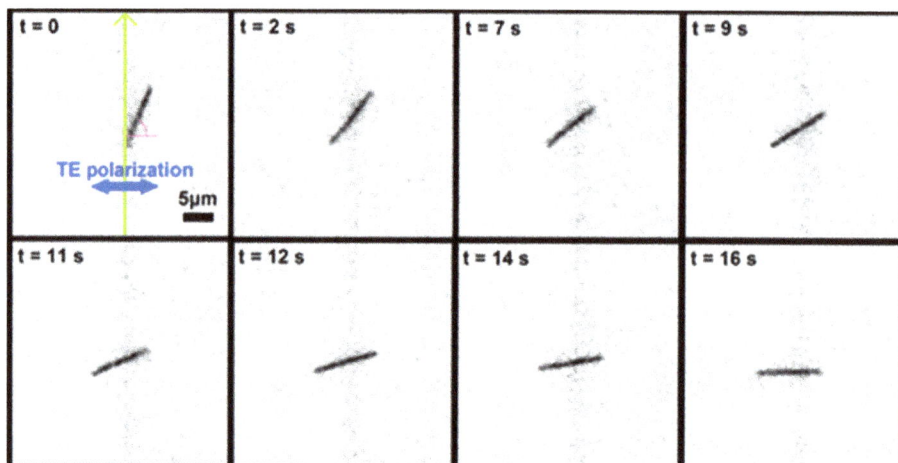

Figure 5.3. Sequential images showing the orientation of a nanorod (black line) into the direction of the electric field of the trapping beam (represented by the blue arrow in the $t = 0$ image); the green arrow denotes the propagation direction of the laser beam. The length of the nanorod is estimated as 10 μm and its diameter is 25 nm. Reprinted with permission from [32]. Copyright 2012 American Chemical Society.

inhomogeneous electromagnetic field—most other nanostructures cannot be satisfactorily modelled in this way. Computational methods are used instead, such as: the T-matrix method [37–41], where non-spherical objects are modelled as clusters of small spheres; or the discrete dipole approximation [42], where the nanoparticle is treated as a collection of dipoles. The continued experimental and theoretical advancements in the optical trapping, manipulating and assembling of nanostructures should lead to the production of novel nanoelectronic and nanophotonic devices in the near future. Furthermore, *laser refrigeration* has been demonstrated, in which an optically trapped nanocrystal immersed in water generates localised cooling (more than 10 °C below ambient conditions) through anti-Stokes photoluminescence [43].

Metal nanoparticles. In the early days of optical trapping it was suggested that metallic objects could not be tweezed, since particles with high reflectivity are pushed out of the high-intensity region of a beam [44]. In 1994, however, Svoboda and Block experimentally showed that metallic Rayleigh particles can act in a manner analogous to their dielectric counterparts when optically trapped [45]—in fact, the larger polarisability of metal implies that the trapping forces will be greater. This finding spurred on progress in the optical manipulation of metal nanoparticles in liquids [17, 46], and even optical control in air has recently been achieved [47]. In contrast to standard non-resonant optical trapping, techniques have been developed based on the resonant excitation of free electrons in metal nanoparticles [48–51]; the oscillation in electron density that results is known as a *plasmon*. When the frequency of the trapping light matches the (usually gold or silver) nanoparticle's plasmon resonance, the optical response properties of the nanoscale metal object become significantly enhanced. Experimental exploration of the trapping of metal

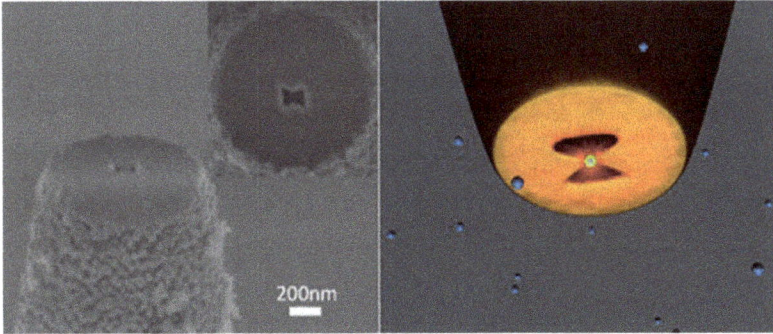

Figure 5.4. Electron beam microscopy image (left) shows the extremity of a plasmon nano-tweezer tip. The sketch on the right exhibits the trapping of a nanoparticle in a 'bowtie' plasmonic aperture. Reproduced by permission from reference [55].

nanoparticles close to their resonance gained momentum following works by Pelton *et al* [52, 53] and Selhuber-Unkel *et al* [54], who optically trapped individual gold nanorods. Yet, despite the steady increase in research on the subject, the field remains in its relative infancy compared to other areas of nanophotonics.

A related plasmonic scheme involves the trapping of nanoparticles within the gaps between plasmonic dipole antennas (a pair of metal nanorods spaced by a nanoscale gap); here it is the localised, evanescent near-field of the nanoantennas that enables the trapping of particles with dimensions as little as 10 nm [56–59]. Capitalising on this advance, nano-optical tweezers capable of three-dimensional manipulation of sub-100 nm dielectric objects are now possible. These traps are built by engineering a bowtie plasmonic aperture at the extremity of an optical fibre (figure 5.4) [55].

Separation of chiral molecules

Beyond the ability of simple detection using microwave spectroscopy [60], recent theory reveals a potential for dielectric nanoparticles in the form of chiral molecules to be separated within an optical trap [61–65]. Chiral molecules have two forms, called enantiomers, that are non-superimposable mirror-images of each other; each enantiomer is readily identifiable by its rotation of linearly polarised light in opposite directions. While the gradient force of optical trapping is usually described within the electric dipole approximation, the mechanism for chiral studies requires that the magnetic transition dipole moments of the molecules are also considered. When the trapping laser is circularly polarised, the optical force that acts on one enantiomer differs slightly to its opposite enantiomer—so that an optical separation is possible. Such a difference in force is very small (with estimates in the 10^{-16} N range for a trapping beam intensity of 5×10^{11} W cm^{-2}) but considered experimentally distinguishable. A potential physical system is given by figure 5.5. An alternative separation scheme, based on an optical trap that uses a plasmonic aperture and

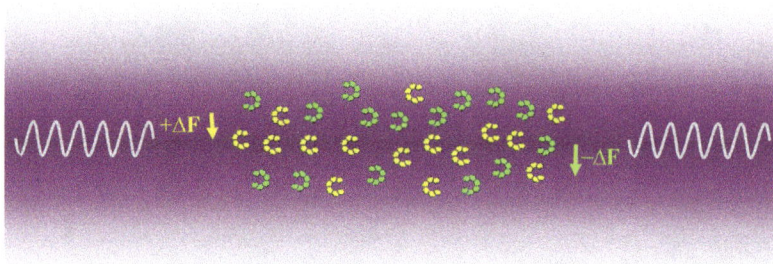

Figure 5.5. Diagram depicting the relative positions, within a circularly polarised trapping beam of Gaussian profile, of two typical enantiomers (S)-hexahelicene (yellow) and (R)-hexahelicene (green). Left-handed (S) molecules have a greater tendency, compared to the right-handed ones (R), to be positioned towards the centre of the Gaussian beam, as denoted by the chiral force ΔF; grey wavy line denotes the throughput beam.

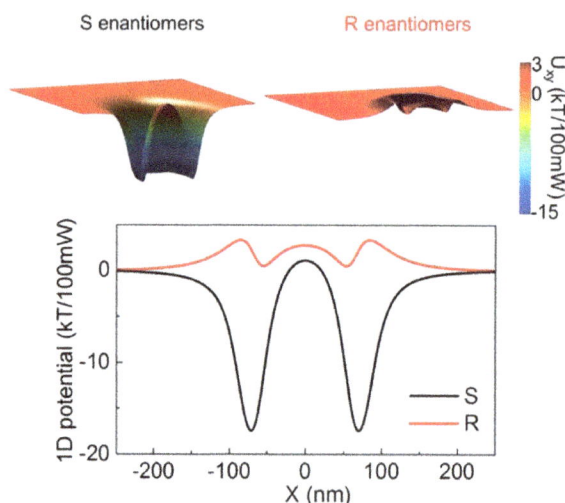

Figure 5.6. Two- and one-dimensional trapping potentials for enantiomers irradiated with circularly polarised light. The energy potential varies across the plasmonic aperture; the case where the R- and S-enantiomers are positioned 20 nm above the aperture are shown. The large difference in the potential well of an S-enantiomer (black curve) compared to an R-enantiomer (red curve) is the source of the selective enantiomer trap. Reproduced by permission from reference [66]. Copyright 2016 American Chemical Society.

circularly polarised light, has been proposed by Zhao *et al*: one enantiomer is trapped in a deep potential well and the other is repelled with a potential barrier (figure 5.6) [66].

References

[1] Ashkin A 1978 Trapping of atoms by resonance radiation pressure *Phys. Rev. Lett.* **40** 729–32
[2] Ashkin A 1980 Applications of laser radiation pressure *Science* **210** 1081–8
[3] Ashkin A and Dziedzic J M 1985 Observation of radiation-pressure trapping of particles by alternating light beams *Phys. Rev. Lett.* **54** 1245–8

[4] Ashkin A and Gordon J P 1983 Stability of radiation-pressure particle traps: an optical Earnshaw theorem *Opt. Lett.* **8** 511–3

[5] Ashkin A, Dziedzic J M, Bjorkholm J E and Chu S 1986 Observation of a single-beam gradient force optical trap for dielectric particles *Opt. Lett.* **11** 288–90

[6] Harada Y and Asakura T 1996 Radiation forces on a dielectric sphere in the Rayleigh scattering regime *Opt. Commun.* **124** 529–41

[7] Malagnino N, Pesce G, Sasso A and Arimondo E 2002 Measurements of trapping efficiency and stiffness in optical tweezers *Opt. Commun.* **214** 15–24

[8] Ashkin A and Dziedzic J M 1987 Optical trapping and manipulation of viruses and bacteria *Science* **235** 1517–20

[9] Ashkin A, Dziedzic J M and Yamane T 1987 Optical trapping and manipulation of single cells using infrared laser beams *Nature* **330** 769–71

[10] Zhong M-C, Wei X-B, Zhou J-H, Wang Z-Q and Li Y-M 2013 Trapping red blood cells in living animals using optical tweezers *Nat. Commun.* **4** 1768

[11] Agate B, Brown C T A, Sibbett W and Dholakia K 2004 Femtosecond optical tweezers for *in situ* control of two-photon fluorescence *Opt. Express* **12** 3011–7

[12] Villangca M J, Palima D, Banas A R and Gluckstad J 2016 Light-driven micro-tool equipped with a syringe function *Light Sci. Appl.* **5** e16148

[13] Ashkin A 1970 Acceleration and trapping of particles by radiation pressure *Phys. Rev. Lett.* **24** 156–9

[14] Constable A, Kim J, Mervis J, Zarinetchi F and Prentiss M 1993 Demonstration of a fiber-optical light-force trap *Opt. Lett.* **18** 1867–9

[15] Zemánek P, Jonáš A, Šrámek L and Liška M 1999 Optical trapping of nanoparticles and microparticles by a Gaussian standing wave *Opt. Lett.* **24** 1448–50

[16] Maragò O M, Jones P H, Gucciardi P G, Volpe G and Ferrari A C 2013 Optical trapping and manipulation of nanostructures *Nat. Nanotechnol.* **8** 807–19

[17] Dienerowitz M, Mazilu M and Dholakia K 2008 Optical manipulation of nanoparticles: a review *J. Nanophoton* **2** 021875

[18] Čižmár T, Dávila Romero L C, Dholakia K and Andrews D L 2010 Multiple optical trapping and binding: New routes to self-assembly *J. Phys. B: At. Mol. Opt. Phys.* **43** 102001

[19] Spesyvtseva S E S and Dholakia K 2016 Trapping in a material world *ACS Photonics* **3** 719–36

[20] Tan S, Lopez H A, Cai C W and Zhang Y 2004 Optical trapping of single-walled carbon nanotubes *Nano Lett.* **4** 1415–9

[21] Plewa J, Tanner E, Mueth D M and Grier D G 2004 Processing carbon nanotubes with holographic optical tweezers *Opt. Express* **12** 1978–81

[22] Maragò O M *et al* 2008 Optical trapping of carbon nanotubes *Physica* E **40** 2347–51

[23] Mishra A, Clayton K, Velasco V, Williams S J and Wereley S T 2016 Dynamic optoelectric trapping and deposition of multiwalled carbon nanotubes *Microsys. Nanoeng.* **2** 16005

[24] Agarwal R, Ladavac K, Roichman Y, Yu G, Lieber C M and Grier D G 2005 Manipulation and assembly of nanowires with holographic optical traps *Opt. Express* **13** 8906–12

[25] Pauzauskie P J, Radenovic A, Trepagnier E, Shroff H, Yang P and Liphardt J 2006 Optical trapping and integration of semiconductor nanowire assemblies in water *Nat. Mater.* **5** 97–101

[26] van der Horst A, Campbell A I, van Vugt L K, Vanmaekelbergh D A M, Dogterom M and van Blaaderen A 2007 Manipulating metal-oxide nanowires using counter-propagating optical line tweezers *Opt. Express* **15** 11629–39

[27] Irrera A, Artoni P, Saija R, Gucciardi P G, Iatì M A, Borghese F, Denti P, Iacona F, Priolo F and Maragò O M 2011 Size-scaling in optical trapping of silicon nanowires *Nano Lett.* **11** 4879–84

[28] Reece P J, Toe W J, Wang F, Paiman S, Gao Q, Tan H H and Jagadish C 2011 Characterization of semiconductor nanowires using optical tweezers *Nano Lett.* **11** 2375–81

[29] Wang F, Toe W J, Lee W M, McGloin D, Gao Q, Tan H H, Jagadish C and Reece P J 2013 Resolving stable axial trapping points of nanowires in an optical tweezers using photoluminescence mapping *Nano Lett.* **13** 1185–91

[30] Roder P B, Manandhar S, Smith B E, Zhou X, Shutthanandan V S and Pauzauskie P J 2015 Photothermal superheating of water with ion-implanted silicon nanowires *Adv. Opt. Mater.* **3** 1362–7

[31] Yu T, Cheong F-C and Sow C-H 2004 The manipulation and assembly of CuO nanorods with line optical tweezers *Nanotechnology* **15** 1732–36

[32] Kang P, Serey X, Chen Y-F and Erickson D 2012 Angular orientation of nanorods using nanophotonic tweezers *Nano Lett.* **12** 6400–7

[33] Maragò O M *et al* 2010 Brownian motion of graphene *ACS Nano* **4** 7515–23

[34] Twombly C W, Evans J S and Smalyukh I I 2013 Optical manipulation of self-aligned graphene flakes in liquid crystals *Opt. Express* **21** 1324–34

[35] Jauffred L, Richardson A C and Oddershede L B 2008 Three-dimensional optical control of individual quantum dots *Nano Lett.* **8** 3376–80

[36] Chen Y-F, Serey X, Sarkar R, Chen P and Erickson D 2012 Controlled photonic manipulation of proteins and other nanomaterials *Nano Lett.* **12** 1633–7

[37] Simpson S H and Hanna S 2007 Optical trapping of spheroidal particles in Gaussian beams *J. Opt. Soc. Am.* A **24** 430–43

[38] Nieminen T A, Rubinsztein-Dunlop H and Heckenberg N R 2001 Calculation and optical measurement of laser trapping forces on non-spherical particles *J. Quant. Spectrosc. Radiat. Transfer* **70** 627–37

[39] Borghese F, Denti P, Saija R and Iatì M A 2007 Optical trapping of nonspherical particles in the T-matrix formalism *Opt. Express* **15** 11984–98

[40] Nieminen T A, Loke V L Y, Stilgoe A B, Knöner G, Brańczyk A M, Heckenberg N R and Rubinsztein-Dunlop H 2007 Optical tweezers computational toolbox *J. Opt. A: Pure Appl. Opt.* **9** S196-203

[41] Borghese F, Denti P, Saija R, Iatì M A and Maragò O M 2008 Radiation torque and force on optically trapped linear nanostructures *Phys. Rev. Lett.* **100** 163903

[42] Draine B T and Flatau P J 1994 Discrete-dipole approximation for scattering calculations *J. Opt. Soc. Am.* A **11** 1491–9

[43] Roder P B, Smith B E, Zhou X, Crane M J and Pauzauskie P J 2015 Laser refrigeration of hydrothermal nanocrystals in physiological media *Proc. Natl Acad. Sci. USA* **112** 15024–9

[44] Ashkin A 1980 Applications of laser radiation pressure *Science* **210** 1081–8

[45] Svoboda K and Block S M 1994 Optical trapping of metallic Rayleigh particles *Opt. Lett.* **19** 930–2

[46] Bendix P M, Jauffred L, Norregaard K and Oddershede L B 2014 Optical trapping of nanoparticles and quantum dots *IEEE J. Sel. Top. Quant. Electron* **20** 15–26

[47] Jauffred L, Taheri S M-R, Schmitt R, Linke H and Oddershede L B 2015 Optical trapping of gold nanoparticles in air *Nano Lett.* **15** 4713–9

[48] Yan Z, Jureller J E, Sweet J, Guffey M J, Pelton M and Scherer N F 2012 Three-dimensional optical trapping and manipulation of single silver nanowires *Nano Lett.* **12** 5155–61

[49] Urban A S, Carretero-Palacios S, Lutich A A, Lohmuller T, Feldmann J and Jackel F 2014 Optical trapping and manipulation of plasmonic nanoparticles: fundamentals, applications, and perspectives *Nanoscale* **6** 4458–74

[50] Lehmuskero A, Johansson P, Rubinsztein-Dunlop H, Tong L and Käll M 2015 Laser trapping of colloidal metal nanoparticles *ACS Nano* **9** 3453–69

[51] Wang X, Rui G, Gong L, Gu B and Cui Y 2016 Manipulation of resonant metallic nanoparticle using 4Pi focusing system *Opt. Express* **24** 24143–52

[52] Pelton M, Liu M, Kim H Y, Smith G, Guyot-Sionnest P and Scherer N F 2006 Optical trapping and alignment of single gold nanorods by using plasmon resonances *Opt. Lett.* **31** 2075–7

[53] Toussaint K C, Liu M, Pelton M, Pesic J, Guffey M J, Guyot-Sionnest P and Scherer N F 2007 Plasmon resonance-based optical trapping of single and multiple Au nanoparticles *Opt. Express* **15** 12017–29

[54] Selhuber-Unkel C, Zins I, Schubert O, Sönnichsen C and Oddershede L B 2008 Quantitative optical trapping of single gold nanorods *Nano Lett.* **8** 2998–3003

[55] Berthelot J, Acimovic S S, Juan M L, Kreuzer M P, Renger J and Quidant R 2014 Three-dimensional manipulation with scanning near-field optical nanotweezers *Nat. Nanotechnol.* **9** 295–9

[56] Grigorenko A N, Roberts N W, Dickinson M R and Zhang Y 2008 Nanometric optical tweezers based on nanostructured substrates *Nat. Photonics* **2** 365–70

[57] Righini M, Ghenuche P, Cherukulappurath S, Myroshnychenko V, García de Abajo F J and Quidant R 2009 Nano-optical trapping of Rayleigh particles and *Escherichia coli* bacteria with resonant optical antennas *Nano Lett.* **9** 3387–91

[58] Zhang W, Huang L, Santschi C and Martin O J F 2010 Trapping and sensing 10 nm metal nanoparticles using plasmonic dipole antennas *Nano Lett.* **10** 1006–11

[59] Juan M L, Righini M and Quidant R 2011 Plasmon nano-optical tweezers *Nat. Photonics* **5** 349–56

[60] Patterson D, Schnell M and Doyle J M 2013 Enantiomer-specific detection of chiral molecules via microwave spectroscopy *Nature* **497** 475–7

[61] Bradshaw D S and Andrews D L 2014 Chiral discrimination in optical trapping and manipulation *New J. Phys.* **16** 103021

[62] Bradshaw D S and Andrews D L 2015 Laser optical separation of chiral molecules *Opt. Lett.* **40** 677–80

[63] Bradshaw D S and Andrews D L 2015 Electromagnetic trapping of chiral molecules: Orientational effects of the irradiating beam *J. Opt. Soc. Am.* B **32** B25-31

[64] Bradshaw D S, Forbes K A, Leeder J M and Andrews D L 2015 Chirality in optical trapping and optical binding *Photonics* **2** 483–97

[65] Bradshaw D S, Leeder J M, Coles M M and Andrews D L 2015 Signatures of material and optical chirality: Origins and measures *Chem. Phys. Lett.* **626** 106–10

[66] Zhao Y, Saleh A A E and Dionne J A 2016 Enantioselective optical trapping of chiral nanoparticles with plasmonic tweezers *ACS Photonics* **3** 304–9

Chapter 6

Larger nanoparticles: Lorenz–Mie regime and beyond

For large nanoparticles, where the diameter of the particle is similar to the wavelength of the trapping light $(a \sim \lambda)$ the generalised Lorenz–Mie formalism should be applied [1, 2]. This classical electrodynamical representation is an extension of the Mie scattering theory, and deals with the interaction between arbitrary shaped beams and particles with symmetry [3, 4]. However, particle trapping in the *Lorenz–Mie regime* is difficult to model [5]. Moreover, the particle sizes that still dominate optical trapping studies are those in the micron range; such microparticles are usually modelled in the *ray optics regime* $(a \gg \lambda)$. Although optical trapping of particles beyond the nanoscale might be considered outside the remit of this book, some of the interesting research that occurs in the Lorenz–Mie and ray optics regimes are, for completeness, briefly discussed in this chapter.

Mie scattering

In larger, transparent particles the trapping light undergoes internal refraction, which leads to an optical force that acts towards the beam centre. This mechanism is another example of the radiation force, since the optical field can be described in terms of rays that transfer linear momentum to the particle following Mie (inelastic) scattering. This notion is known as the ray optics approach, which is shown by figure 6.1, and is described as follows. The total beam can be decomposed into individual rays—each with an appropriate intensity, direction and polarisation—that propagates in straight lines within uniform media of refractive index n_ω. Each ray may change direction when it reflects or refracts, and it changes polarisation at dielectric interfaces according to the Fresnel equations (which describe the behaviour of light when it traverses the interface between media of different refractive indices). The refraction of light rays at the particle surfaces results in a transfer of momentum from the trapping

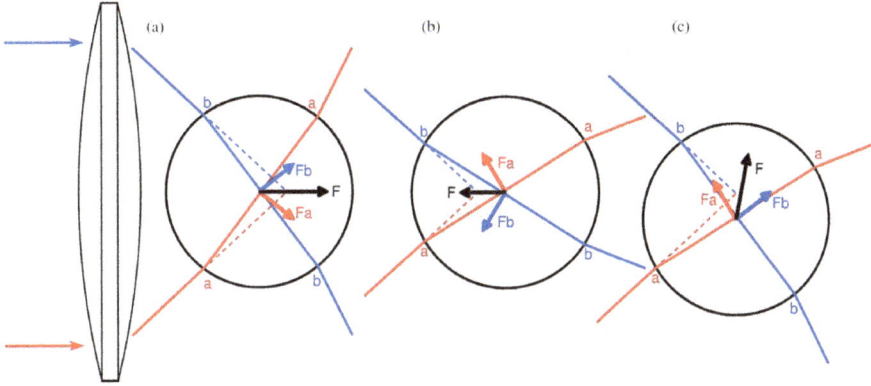

Figure 6.1. Qualitative view of the trapping of a dielectric sphere, showing that the refraction of a typical pair of rays a and b of the trapping beam result in forces, F_a and F_b respectively, whose sum is always restoring towards equilibrium for axial and transverse displacements of the sphere from the trap focus, located at the intersection of the dashed lines. Cases (a) and (b) show the axial restoring force acting on a particle that is axially displaced from the focal plane. Similarly, (c) shows the lateral restoring force upon the particle towards the beam focus. Reproduced by permission from reference [6].

laser to the particle; the rate of delivery determines the overall trapping force [6, 7]. In quantum terms, these larger particles experience a propelling force towards the beam centre through the cumulative result of the directional differences in momentum between each photon entering and exiting the particle. Once again, it is the overall rate of such scattering events that determines the magnitude of the optical force.

Thermal effects on a trapped microparticle

In thermal equilibrium, the natural Brownian motion of an optically trapped microparticle is confined to a finite trapping region. The probability density, N', for a particle position within the trap can be established from Boltzmann statistics, so that in one dimension;

$$N' \propto e^{-\frac{\Delta E(x)}{k_B T}}, \tag{6.1}$$

where T represents the absolute temperature. It is interesting that N' tends to a position-independent constant for high temperatures or low potential energies, which physically represents a loss of a viable optical trap. Following a Taylor series expansion of equation (6.1), the leading term for a local position within the trap is given as;

$$\Delta N' \propto -\frac{\Delta E(x)}{k_B T}, \tag{6.2}$$

which is analogous to Curie's law. It is usual for the trapping potential to be modelled as a harmonic potential, meaning that $\Delta E(x) \sim \frac{1}{2}k_x x^2$ where k_x is a force constant that relates to the stiffness of the trap. The latter can be expressed as

$k_x \sim k_B T / \sigma_x^2$ with σ_x as the variance of a Gaussian distribution curve. Inserting the last two formulae into equation (6.2), the following is determined;

$$\Delta N' \propto -\frac{x^2}{2\sigma_x^2}. \tag{6.3}$$

Here, the drag due to the viscosity of the medium is not considered; this would require additional terms that are detailed elsewhere [8]. On taking a statistically significant number of position measurements, the histogram of particle positions (typically of the form shown in figure 6.2) has a similar shape to the spatial probability density of equation (6.1). As expected, the position of the particles is most probable at the centre of the beam where the potential energy is lowest. These three expressions equally apply to optical trapped nanoparticles. However, particles of such size have smaller polarisabilities (and, hence, smaller ΔE), which results in less variation in the probability density, and a shallower optical trap.

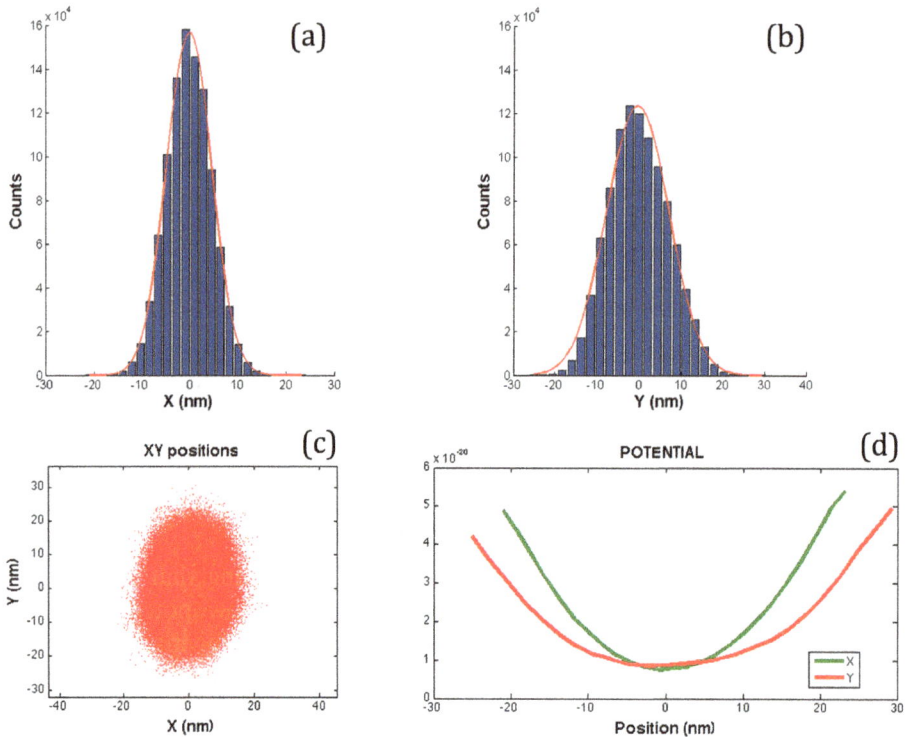

Figure 6.2. Histograms of (a) x-axis position and (b) y-axis position of a particle within an optical trap, and (c) a scatterplot of the particle position within the x, y-plane; all are determined by a repeated series of position measurements. (d) Potential energy curve (in Joules) for the optical trap in the x- and y-directions. The ellipsoid shape in (c) is due to disparity between the potential wells and corresponding particle distributions in the x- and y-directions. Images courtesy of Mario Montes Usategui and Estela Martín Badosa, Universitat de Barcelona.

Developments in microparticle trapping

The subject of optical trapping and manipulation of microparticles contains a colossal amount of material, the coverage of which appears in numerous review articles [5, 9–17]. This section offers a very brief outline on the topic, ranging from the observation of optical levitation in the 1970s to the latest research on optical tweezing in microbiology.

Optical levitation. Developed by Ashkin in the earlier years of laser trapping, optical levitation is a technique whereby a microparticle is suspended in air (or a high vacuum [18]) when the force of gravity, i.e. the particle mass m multiplied by the acceleration due to gravity g, is opposed by the radiation force of a vertically directed focused beam. The Maxwell–Bartoli formula of equation (3.1), which is generally geared towards large scale illumination of an inclined plane, can be readily modified to account for the realities of optical levitation. Assuming a particle of almost spherical shape, and dimensions smaller than the focused beam width, we interpret the light irradiance acting on the under surface of the particle as being only a fraction of the beam intensity. Moreover, a reflective microparticle will reflect light at various angles according to the positions at which each photon hits the surface, so most of the light is not reflected back towards its source. Assuming that the majority of the laser intensity at the focus lies within the beam-width parameter w_0, and for simplicity neglecting the variation of intensity across that region, by use of simple geometry the following force equation is found;

$$mg = \left(\frac{W}{c}\right)\left(\frac{a}{2w_0}\right)^2\left(1 + \frac{R}{3}\right), \tag{6.4}$$

where R is the reflectivity of the levitated particle and W is the power of the beam—averaged over time, if a rapidly pulsed source is deployed. This expression can be re-arranged to produce the following;

$$W = \frac{4\pi r w_0^2 (\rho - \rho_0) c g}{R + 3}. \tag{6.5}$$

Here, $\rho - \rho_0$ is the difference between the density of the particle and a supporting medium (as commonly applied to particles in liquid suspension). For example, using the values $2w_0 = 25$ μm, $\rho - \rho_0 = 50$ kg m^{-3} and $R = 0.6$, it emerges that a reflective particle with a diameter of 20 μm can be levitated by a laser with a power greater than 0.8 mW. Once aloft, the microparticle can be freely manipulated by simply moving the beam. However, unlike optical tweezers, this technique cannot be considered all-optical since it requires gravitational forces to function. Levitation of this type was initially demonstrated with a small transparent glass sphere [19] (figure 6.3) and, soon after, it was also achieved with oil and water droplets [20]. In more recent times, the optical levitation of a microdroplet containing a single quantum dot [21], and the suspension of a microscopic particle within a Bessel (non-diffracting) light beam [22], have been demonstrated. Furthermore, it has been proposed that optically levitating a small curved mirror will enable its mechanically

Figure 6.3. Photograph by Arthur Ashkin of an optically levitated glass sphere in air. Reproduced with permission from [9]. Copyright 2006 World Scientific.

isolated, non-contact support to be achieved, and thereby utilised for ultrasensitive motion detection [23].

Trapping of microbiological particles. For most applications, optical tweezer set-ups are far superior to levitation techniques. Although originally designed as a trap for atoms, the optical tweezing of microbiological particles quickly gained favour. It is easy to understand why: the drawback of much greater mass can be compensated by suspension in a liquid (usually aqueous) medium, while the much larger surface area affords a far more effective response to the laser light. Moreover, the relatively slow rates of particle diffusion for sub-micron (compared to atomic scale) particles in the liquid provide for more easily controlled manipulation. The earliest tweezers could trap individual tobacco mosaic viruses (a rod-like protein) and *Escherichia coli* bacteria [24]; by avoiding any optical damage, even the process of cell reproduction of the latter can be observed within the trap [25]. To circumvent any photodestruction of biological particles over time, a near-infrared wavelength range of 750–1100 nm (conveniently produced by a titanium:sapphire laser) is usually chosen for the trapping beam rather than visible light—most biological samples are transparent in this range, so that damage due to light absorption is precluded. However, cell impairment through multiphoton absorption of near-infrared light has been reported [26]. Today, optical tweezing is no longer restricted

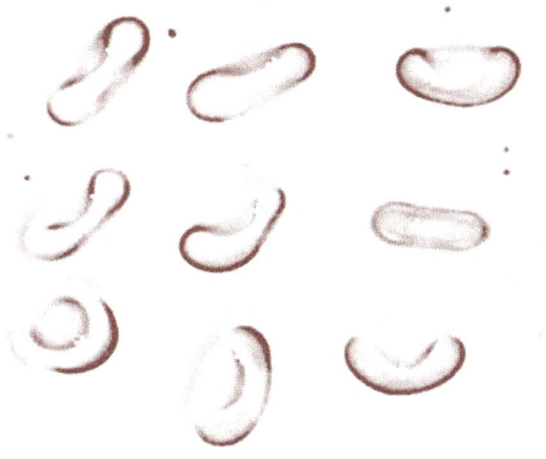

Figure 6.4. Multiplexed optical trapping of human red blood cells in a 3 × 3 array produced with an acousto-optic modulator. Each cell is approximately 6–8 microns in diameter. The laser operates at 1064 nm with 900 mW output power. Image courtesy of Kishan Dholakia, University of St Andrews.

to the trapping of single microbiological particles. Conventional single-beam tweezing methods have been replaced by multiple traps [27]—for example, red blood cells can now be trapped in an optical array (figure 6.4)—and techniques that use laser beams with exotic forms of wavefront structure [28]. Optical trapping arrays are discussed in the next chapter, while manipulation using structured light beams is detailed in chapters 8 and 9.

Force measurements in biology. Another important application of optical tweezers is the study of molecular motors. These mechanoenzymes interact with the microtubules or actin filaments within the cell to produce forces that are responsible for movement: examples include cell motility and muscle action. Before the advent of optical tweezers, the understanding of biomolecular processes was limited by the fact that only bulk biological samples could be examined. Tweezing enabled unprecedented insight into the interactions of biological motors (and other biopolymers) at the molecular level. For instance, Block *et al* observed the motor protein known as kinesin traverse a microtubule in a sequence of 8 nm steps [29] and picoNewton-sized forces were recorded for kinesin motion [30]; an illustration of a kinesin molecule 'walking' along a microtubule is shown in figure 6.5. This study uses the 'handles' technique, in which the microbiological particle is attached to a small dielectric sphere that is optically manipulated. The method is also useful for applying a mechanical force across a biomolecule, as illustrated by figure 6.6, so that its properties (such as elasticity) can be found. For example, on attaching straddle beads (the 'handles') to each end of a DNA strand, the relationship between the applied force and the extension of the DNA can be revealed [31–34] and the

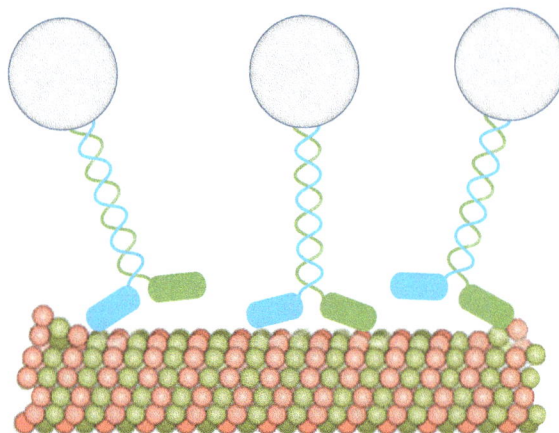

Figure 6.5. Molecular motor kinesin with a silica bead attached, depicted in three different stages of progression along the surface of a microtubule. Using optical tweezers to arrest the bead enables the motive force to be determined.

molecular process of DNA folding can be measured [35]—this is essential knowledge since the flexibility and winding of DNA play a crucial role in cellular functionality. Beyond simple stretching experiments, the fashioning of a knot into a single DNA strand (with a handle at each end) has also been achieved using optical tweezers [36]. The creation of more complex DNA knots, and the analysis of their behaviour, could lead to a better understanding of how DNA remains knot-free inside a cell [37].

Cell sorting. The current industrial standard for sorting a heterogeneous mixture of biological cells in flow cytometry is known as the *fluorescence-activated cell sorter* [38]. In this system, the cells (often attached to a fluorescent marker) are suspended in a rapidly flowing stream of liquid. While traversing the apparatus, the cells are first excited by a laser beam, and the detection of the subsequent characteristic fluorescence is used to identify the cell type. Depending on the latter, the cells are deflected into a specific direction and then collected into the appropriate reservoir.

Despite the success and widespread adoption of such a technique, novel passive separation methods [11] based on optical micromanipulation have emerged that do not require fluorescent markers—since each cell is identified by its physical attributes in this procedure—and these offer a number of key advantages [39]. Firstly, rather than sorting on the bulk scale, cell sorters based on optical trapping are capable of working on a single cell basis, so that the sample size can be significantly reduced. Secondly, passive sorting has a high positional accuracy, since the technique acts on each cell rather than a volume of fluid that contains the cell. Finally, sorting by micromanipulation is a completely sterile, non-contact procedure. To summarise, passive cell sorting requires three main steps: identification of the cell to be sorted, usually by morphological means; trapping of the cell; and manipulation of the cell

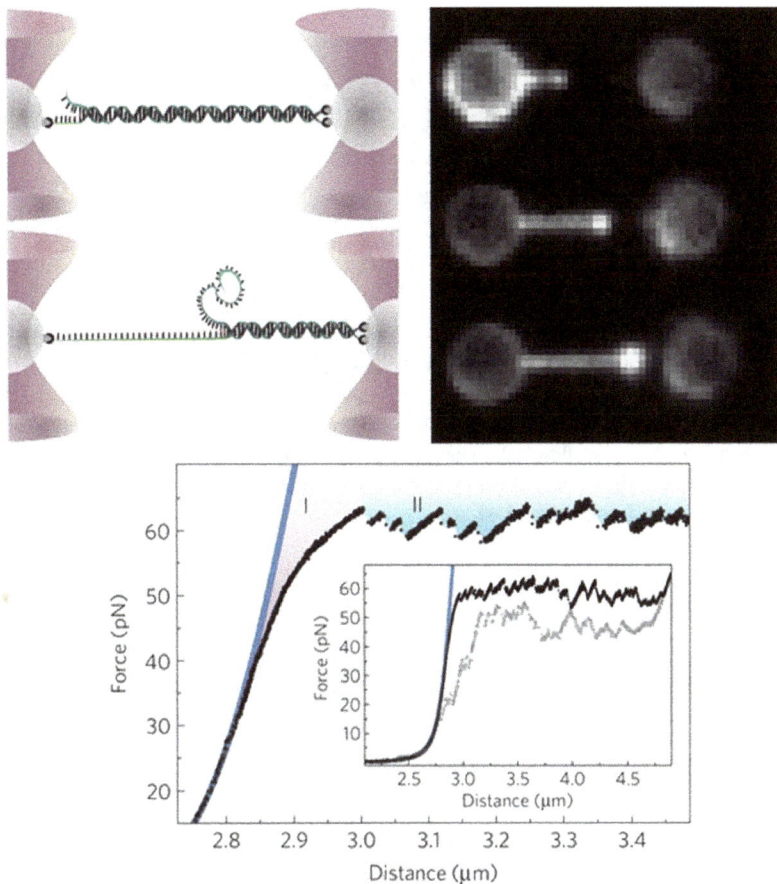

Figure 6.6. Force-induced strand unpeeling of DNA studied with fluorescence microscopy and force spectroscopy. (Top left) Schematic depictions of DNA undergoing unpeeling on extension, where three ends of the DNA strands are connected to the optically trapped microspheres. (Top right) Three fluorescence images of partially overstretched DNA molecules. Only the left part of the DNA construct is fluorescent, indicating that unpeeling occurs only from the left side of the DNA. (Bottom) Force–extension measurement of the DNA construct. Region I: strong deviation from the extensible model (blue line) is evident. Region II: an overstretching transition with pronounced stepwise increases in length. Inset: extended view of the same curve (black) together with the reverse process, in which strand reannealing occurs on relaxation (grey). Reproduced by permission from reference [34].

into a designated reservoir. One way to achieve microfluidic sorting is via an array of optical traps that relates to an intensity pattern with specifically designed spatial variations; this is known as *optical fractionation* (figure 6.7). Optical forces in such patterns depend on the microparticle shape, size and refractive index [40]. Many other techniques have been proposed, both with and without the use of a fluidic flow [41–47].

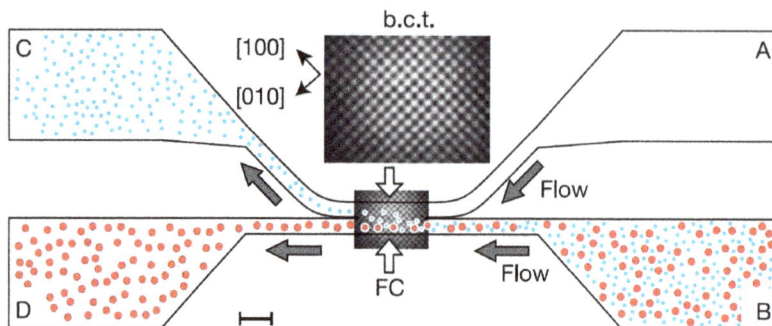

Figure 6.7. The concept of optical fractionation. Without an actuator all particles from chamber B will flow into chamber D, as a result of the non-turbulence of microfluidic flow. On introduction of an optical lattice—in this case, a body-centred tetragonal (b.c.t.) lattice—into the fractionation chamber (FC), one type of particle is selectively pushed into the upper flow field; chamber A relates to a 'blank' flow stream and the scale bar represents 40 μm. The reconfigurability of the optical lattice allows for dynamic updating of selection criteria. Reproduced by permission from reference 40.

References

[1] Harada Y and Asakura T 1996 Radiation forces on a dielectric sphere in the Rayleigh scattering regime *Opt. Commun.* **124** 529–41

[2] Malagnino N, Pesce G, Sasso A and Arimondo E 2002 Measurements of trapping efficiency and stiffness in optical tweezers *Opt. Commun.* **214** 15–24

[3] Gouesbet G 2009 Generalized Lorenz–Mie theories, the third decade: A perspective *J. Quant. Spectrosc. Radiat. Transfer* **110** 1223–38

[4] Gouesbet G 2014 Latest achievements in generalized Lorenz–Mie theories: A commented reference database *Ann. Phys.* (Berlin) **526** 461–89

[5] Dholakia K, Reece P and Gu M 2008 Optical micromanipulation *Chem. Soc. Rev.* **37** 42–55

[6] Čižmár T, Dávila Romero L C, Dholakia K and Andrews D L 2010 Multiple optical trapping and binding: New routes to self-assembly *J. Phys. B: At. Mol. Opt. Phys.* **43** 102001

[7] Ashkin A 1992 Forces of a single-beam gradient laser trap on a dielectric sphere in the ray optics regime *Biophys. J.* **61** 569–82

[8] Mas J, Farre A, Cuadros J, Juvells I and Carnicer A 2011 Understanding optical trapping phenomena: A simulation for undergraduates *IEEE Trans. Educ.* **54** 133–40

[9] Ashkin A 1997 Optical trapping and manipulation of neutral particles using lasers *Proc. Natl. Acad. Sci. USA* **94** 4853–60

[10] Dholakia K and Reece P 2006 Optical micromanipulation takes hold *Nano Today* **1** 18–27

[11] Jonáš A and Zemánek P 2008 Light at work: The use of optical forces for particle manipulation, sorting, and analysis *Electrophoresis* **29** 4813–51

[12] Zhang H and Liu K-K 2008 Optical tweezers for single cells *J. R. Soc. Interface* **5** 671–90

[13] Stevenson D J, Gunn-Moore F and Dholakia K 2010 Light forces the pace: optical manipulation for biophotonics *J. Biomed. Opt.* **15** 041503

[14] Verdeny I, Farré A, Mas J, López-Quesada C, Martín-Badosa E and Montes-Usategui M 2011 Optical trapping: a review of essential concepts *Opt. Pura Apl.* **44** 527–51

[15] Bowman R W and Padgett M J 2013 Optical trapping and binding *Rep. Prog. Phys.* **76** 026401

[16] Palima D and Glückstad J 2013 Gearing up for optical microrobotics: micromanipulation and actuation of synthetic microstructures by optical forces *Laser & Photon. Rev.* **7** 478–94

[17] Daly M, Sergides M and Nic Chormaic S 2015 Optical trapping and manipulation of micrometer and submicrometer particles *Laser & Photon. Rev.* **9** 309–29

[18] Ashkin A and Dziedzic J M 1976 Optical levitation in high vacuum *Appl. Phys. Lett.* **28** 333–5

[19] Ashkin A and Dziedzic J M 1971 Optical levitation by radiation pressure *Appl. Phys. Lett.* **19** 283–5

[20] Ashkin A and Dziedzic J M 1975 Optical levitation of liquid drops by radiation pressure *Science* **187** 1073–5

[21] Minowa Y, Kawai R and Ashida M 2015 Optical levitation of a microdroplet containing a single quantum dot *Opt. Lett.* **40** 906–9

[22] Garcés-Chávez V, Roskey D, Summers M D, Melville H, McGloin D, Wright E M and Dholakia K 2004 Optical levitation in a Bessel light beam *Appl. Phys. Lett.* **85** 4001–3

[23] Guccione G, Hosseini M, Adlong S, Johnsson M T, Hope J, Buchler B C and Lam P K 2013 Scattering-free optical levitation of a cavity mirror *Phys. Rev. Lett.* **111** 183001

[24] Ashkin A and Dziedzic J M 1987 Optical trapping and manipulation of viruses and bacteria *Science* **235** 1517–20

[25] Ashkin A, Dziedzic J M and Yamane T 1987 Optical trapping and manipulation of single cells using infrared laser beams *Nature* **330** 769–71

[26] König K, Liang H, Berns M W and Tromberg B J 1996 Cell damage in near-infrared multimode optical traps as a result of multiphoton absorption *Opt. Lett.* **21** 1090–2

[27] Chiou P Y, Ohta A T and Wu M C 2005 Massively parallel manipulation of single cells and microparticles using optical images *Nature* **436** 370–2

[28] Arlt J, Garces-Chavez V, Sibbett W and Dholakia K 2001 Optical micromanipulation using a Bessel light beam *Opt. Commun.* **197** 239–45

[29] Svoboda K, Schmidt C F, Schnapp B J and Block S M 1993 Direct observation of kinesin stepping by optical trapping interferometry *Nature* **365** 721–7

[30] Svoboda K and Block S M 1994 Force and velocity measured for single kinesin molecules *Cell* **77** 773–84

[31] Gross P, Laurens N, Oddershede L B, Bockelmann U, Peterman E J G and Wuite G J L 2011 Quantifying how DNA stretches, melts and changes twist under tension *Nat. Phys.* **7** 731–6

[32] Smith S B, Cui Y and Bustamante C 1996 Overstretching B-DNA: The elastic response of individual double-stranded and single-stranded DNA molecules *Science* **271** 795–9

[33] Wang M D, Yin H, Landick R, Gelles J and Block S M 1997 Stretching DNA with optical tweezers *Biophys. J.* **72** 1335–46

[34] Bustamante C, Bryant Z and Smith S B 2003 Ten years of tension: single-molecule DNA mechanics *Nature* **421** 423–7

[35] Neupane K, Foster D A N, Dee D R, Yu H, Wang F and Woodside M T 2016 Direct observation of transition paths during the folding of proteins and nucleic acids *Science* **352** 239–42

[36] Arai Y, Yasuda R, Akashi K-i, Harada Y, Miyata H, Kinosita K and Itoh H 1999 Tying a molecular knot with optical tweezers *Nature* **399** 446–8

[37] Bao X R, Lee H J and Quake S R 2003 Behavior of complex knots in single DNA molecules *Phys. Rev. Lett.* **91** 265506

[38] Herzenberg L A, Parks D, Sahaf B, Perez O, Roederer M and Herzenberg L A 2002 The history and future of the fluorescence activated cell sorter and flow cytometry: A view from Stanford *Clin. Chem.* **48** 1819–27

[39] Grover S C, Skirtach A G, Gauthier R C and Grover C P 2001 Automated single-cell sorting system based on optical trapping *J. Biomed. Opt.* **6** 14–22

[40] MacDonald M P, Spalding G C and Dholakia K 2003 Microfluidic sorting in an optical lattice *Nature* **426** 421–4

[41] Ladavac K, Kasza K and Grier D G 2004 Sorting mesoscopic objects with periodic potential landscapes: Optical fractionation *Phys. Rev.* E **70** 010901

[42] Paterson L, Papagiakoumou E, Milne G, Garcés-Chávez V, Tatarkova S A, Sibbett W, Gunn-Moore F J, Bryant P E, Riches A C and Dholakia K 2005 Light-induced cell separation in a tailored optical landscape *Appl. Phys. Lett.* **87** 123901

[43] Čižmár T, Šiler M, Šerý M, Zemánek P, Garcés-Chávez V and Dholakia K 2006 Optical sorting and detection of submicrometer objects in a motional standing wave *Phys. Rev.* B **74** 035105

[44] Rodrigo P J, Perch-Nielsen I R and Glückstad J 2006 Three-dimensional forces in GPC-based counterpropagating-beam traps *Opt. Express* **14** 5812–22

[45] Dholakia K, MacDonald M P, Zemánek P and Čižmár T 2007 Cellular and colloidal separation using optical forces *Methods in Cell Biology* ed M Berns and K Greulich (Cambridge, MA: Academic) pp 467–95

[46] Eriksson E, Scrimgeour J, Granéli A, Ramser K, Wellander R, Enger J, Hanstorp D and Goksör M 2007 Optical manipulation and microfluidics for studies of single cell dynamics *J. Opt.* A: *Pure Appl. Opt.* **9** S113–21

[47] Rindorf L, Bu M and Glückstad J 2014 Parallel optical sorting of biological cells using the generalized phase contrast method *Proc. SPIE* **8976** 89760U

Chapter 7

Optical trapping arrays

Optical trapping arrays, alternatively known as optical lattices or 'artificial crystals of light', represent an extremely useful tool amongst modern methods of optical manipulation. Essentially they comprise a set of two-dimensional or three-dimensional traps, originally created by the interference of counter-propagating laser beams (figure 7.1) [1–4]. The resulting optical potential landscape exhibits a periodic pattern. In such a system, micro- and nano-scale particles can be trapped within individual potential wells due to the gradient force (which is described in chapter 3). A variety of properties that characterise the lattice potential can be controlled in real-time—such as the periodicity [5]—through changes in the relative phase, polarisation, frequency and intensity of the laser beams. Historically, the interest in optical trapping arrays stems from the study of the laser cooling of atoms [6]. More recently, for particles at room temperatures, alternative methods for optical lattice production have been successfully implemented based on time-sharing and holographic techniques.

Ultracold atoms: optical lattices and quantum information

Optical lattices can be used to confine ultracold atoms [7–13]. In dealing with the electronic properties of individual atoms, whose quantised energy levels are essentially discrete and relate to a narrow spectral linewidth, the polarisability is generally dominated by the first term of equation (3.8). This is especially so when the optical frequency of the input beam, ω, is similar to that of a specific electronic transition $0 \to s$, where s is the excited state in an assumed two-level system. Then, $E_{s0} \simeq \hbar\omega$, so that $\delta E = E_{s0} - \hbar\omega$ is small. To properly account for the optical response in such near-resonance conditions (i.e. the *dispersion* behaviour), it is necessary to introduce a damping term associated with the finite lifetime of excited

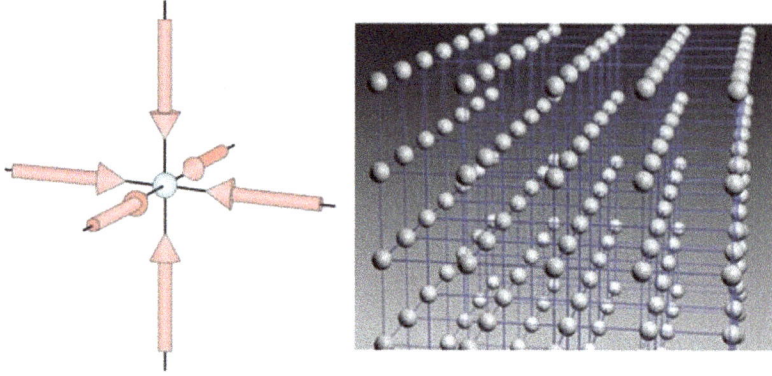

Figure 7.1. Three-dimensional optical lattice formed, at the co-located focus of three pairs of counter-propagating laser beams, through superimposition of the resultant standing waves in each dimension (left). The potential energy surface for such a lattice is approximated by a 3D simple cubic array of tightly confining harmonic oscillator potentials at each lattice site (right). Reproduced by permission from reference [4].

states to the denominator of equation (3.6) [14]. The following expression is then found[1];

$$\alpha_{ij}(\omega) \simeq \frac{\mu_i^{0s}\mu_j^{s0}}{\delta E - i\hbar\Omega_s}, \tag{7.1}$$

where Ω_s is the HWHM (half-width at half-maximum) of a narrow Lorentzian lineshape in the scattering profile. From equation (7.1) it is clear that for red detuning, $\delta E > 0$, the polarisability is positive and the usual gradient force is attractive. Conversely, blue detuning produces a repulsive force, away from the most intense part of the beam. From the same expression, moreover, we see that the optical trapping of atoms involves their transition electric dipole moments. For cases beyond the dipole approximation, optical lattices based on electric quadrupole moments have been considered [15]. The motion of the trapped ultracold atoms is quantised and dependent on their vibrational motion within individual wells and the transfer of atoms between the potential minima (which may involve quantum tunnelling [16]); this is known as Bloch oscillation [17, 18].

The storing of ultracold atomic gases within optical lattices, as shown by figure 7.2, has led to numerous innovative possibilities [4, 19]. One of the most enticing prospects is for ultracold atoms trapped within optical arrays to be utilised as qubits for quantum information processing [20–26]. The possibility arises because ultracold atoms in their ground states couple extremely weakly to both the environment and each other, and as a result quantum decoherence (which leads irreversibly to a transformation of the qubits into classical bits) is suppressed.

---

[1] The introduction of an imaginary damping term into the polarisability formula is essentially pragmatic and phenomenological. This limitation arises since it is impossible to reconcile any rigorous form with the demands of time-reversal invariance, due to the non-Hermitian nature of the Hamiltonian for an implicitly non-conservative system.

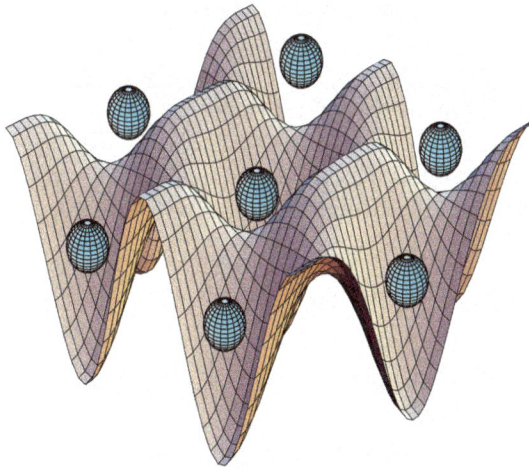

Figure 7.2. Potential energy surface of an optical lattice, schematically identifying the location of trapped atoms. Figure from the Rolston group website (www.physics.umd.edu/rgroups/amo/rolstonwebsite/bec.htm). Reproduced with permission from Professor Rolston, University of Maryland, USA

However, the weak form of interaction between the atoms is also a disadvantage, since a vital requisite for the register of a quantum computer is the interaction known as entanglement [27][2].

'On demand' methods to generate the required entanglement are based on either the application of an auxiliary laser beam to induce dipole–dipole interactions [28, 29] (using a method closely related to the optical binding techniques discussed in chapter 12) or 'cold controlled collisions' [30, 31]—a procedure that enhances local interactions by moving the potential wells closer together or by modifying the shape of the wells. The ability to activate interactions only when required is highly beneficial in overcoming decoherence. Experimental realisation of a quantum computer remains a major challenge. For instance, although some advances have been made [32, 33], a robust and reliable method is still required for addressing and reading the state of the qubits, complicated by atom separations that are generally smaller than an optical wavelength. Requirements such as the stabilisation of the local surroundings, which may encounter rogue electric and magnetic fields, are equally demanding.

Ultracold atoms within an optical lattice can also act as a basis for a high precision optical clock [34–36]. The extraordinarily accurate timekeeping in such a system is due to the narrow spectral linewidth of the trapped atoms, which is insensitive to external perturbations and effects such as the Doppler shift. These atomic clocks (often based on strontium) will neither gain nor lose a second over the estimated age of the Universe [37], and display higher accuracy than the caesium

---

[2] In entangled states, a non-factorisable combination of individual component (atom) wavefunctions means that the quantum state can only be defined with respect to the system as a whole.

clocks used to determine International Atomic Time [38]—so much so that the SI unit of time, seconds, could be redefined in the near future. The reason for the improved precision is that optical lattice clocks measure light emitted in the visible range, rather than lower frequency microwaves, and also utilise a large number of trapped atoms so that signal errors are averaged out. Optical atomic clocks have many potential applications, which range from improved satellite navigation systems to ultrasensitive tests of fundamental physical theories.

Nanoparticles in suspension: techniques for optical lattice production

Time-sharing. When an optically trapped particle is dispersed within a liquid medium (for example, water), escape from the vicinity of the trap may be precluded by viscous forces even if the laser beam is removed for a short time. In such a scenario, any momentarily released particle will have a diffusion rate that conforms to Brownian motion, and the diffusion distance d of a particle for a beam removal time t is determined from;

$$d = \sqrt{\frac{2k_{\mathrm{B}}Tt}{3\pi\eta a}}, \tag{7.2}$$

where η denotes the viscosity of the medium. Hence, smaller particles have a capacity to migrate over larger diffusion distances, when the laser beam is absent. Consider, for example, a particle of 100 nm diameter immersed at room temperature, 298 K, in an aqueous medium ($\eta = 8.9 \times 10^{-4}$ Pa s). Calculation shows that particle translation by a full diameter can be expected in a timescale of approximately one millisecond. Ideally, the optical trap should, therefore, be removed for no longer than around 0.1 ms; indeed this is a typical value. Since the trapping laser can temporarily be moved to other positions (at great speed, for example, using computer-controlled deflectors), it is possible to simultaneously trap multiple particles by rapidly time-sharing the beam between each trapping site. Naturally, this time-sharing technique is limited by a maximum number of optical traps, which clearly depends on the size of the particles. The concept of a scanning laser trapping technique was demonstrated by Sasaki *et al* [39, 40]. Here, particles were aligned into a designated two-dimensional pattern by a single trapping laser that scans repeatedly across a specific route. Such systems, driven by mirror galvanometers, have been developed to achieve time-sharing rates around 1 kHz [41]—more than adequate for particles of micron size, but problematic for anything much smaller.

Currently, the most popular forms of time-shared traps, offering tunable and highly precise beam-steering, are generated by acousto-optic deflectors (AODs)—as presented by Visscher *et al* [42]. The principle is based on a piezoelectric transducer generating an acoustic wave at MHz frequencies inside a transparent crystal, such as tellurium oxide. This is used to induce a standing wave modulation of refractive index along the path of a propagating light beam. The resulting diffraction can then be steered through an angle that is under the control of the signal frequency applied to the transducer. A pair of AODs perpendicularly aligned along the propagation axis provides for the generation of a number of optical traps, arbitrarily displaced over the focal plane. Computer-controlled AODs allow for a flexibility in the number and relative strengths of the optical traps, as well as control of the overall spatial pattern; adjustable time-sharing rates up to 10 kHz are achievable. While this technique is based on multi-particle manipulation within a plane, more elaborate optical schemes may generate additional layers [43]. Outside of systems based on AODs, electro-optic deflectors (EODs) can be used as an alternative scheme for modifying the refractive index gradient of the medium [44]. In such a case, which involves the application of an external electric field, the time-sharing rates are much faster than most systems that use AODs; however, a major downfall is that the attainable deflection angles are very small. Figure 7.3 offers a typical experimental set-up for the generation of multiple optical traps with AODs or EODs as the deflector.

Holography. Instead of time-sharing the trapping beam between numerous sites, it is ideally desirable to distribute the beam between each site simultaneously. A popular technique to achieve such an aim is based on holography [45–47]. Conventionally, the latter involves the splitting of a laser beam into two parts: one is reflected by an object before acting on a recording medium (the illumination beam), the other travels directly to the same photographic film (the reference beam). The interference pattern between the two beams is imprinted onto the recording medium, which is known as the hologram. On shining the original laser onto the developed film, the light is diffracted by the surface pattern of the hologram and an image of the object is generated. Although the concept of digital holography is implemented in holographic optical tweezers (HOTs), the technique does *not* require an actual object and no hologram is physically recorded. The optical trapping array is produced by diffracting laser light with a computer-generated hologram, with an appropriate algorithm fashioning the form of the array. Such a diffractive element (hologram), in its static form, is microfabricated by etching a surface pattern onto fused silica or a polymeric material [48, 49]. The alternative, dynamic diffractive element uses a computer-addressed spatial light modulator (SLM) to impose, in real time, a prescribed amount of phase shift on each liquid crystal pixel in a display

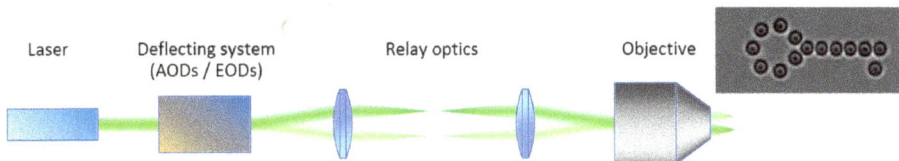

Figure 7.3. A typical experimental set-up for the generation of multiple optical traps using the time-sharing method. Although the laser beam is present at only one trapping position at any instant in time, fast switching between these positions enables up to several hundred time-shared optical traps to be sustained in one axial plane. The beam-steering element (acousto-optic deflectors, AODs, or electro-optic deflectors, EODs) and the objective back apertures are placed at conjugated planes of a relay telescope. Inset: beads held in place by a typical optical lattice generated by this technique. Reproduced by permission from reference [45].

[50–60]. Figure 7.4 schematically portrays a HOTs system that contains a dynamic diffractive element.

Holographic optical tweezers have many advantages over the time-sharing schemes, including the capacity to: create optical traps in three dimensions, shape each trap individually, arbitrarily distribute laser power between the traps and remove any optical aberrations. However, they require a complex computation process for large trapping arrays that is not required by other techniques, such as the generalised phase contrast method [61–64] (which has its own disadvantages, for example, the problem of highly focussing the trapping beams). Nanoscale applications of HOTs include the positioning and prising apart of carbon nanotube bundles [65], and the manipulation and assembly of nanowires [66–68]. The latter is depicted in figure 7.5, which shows the nanofabrication of a rhombus structure using

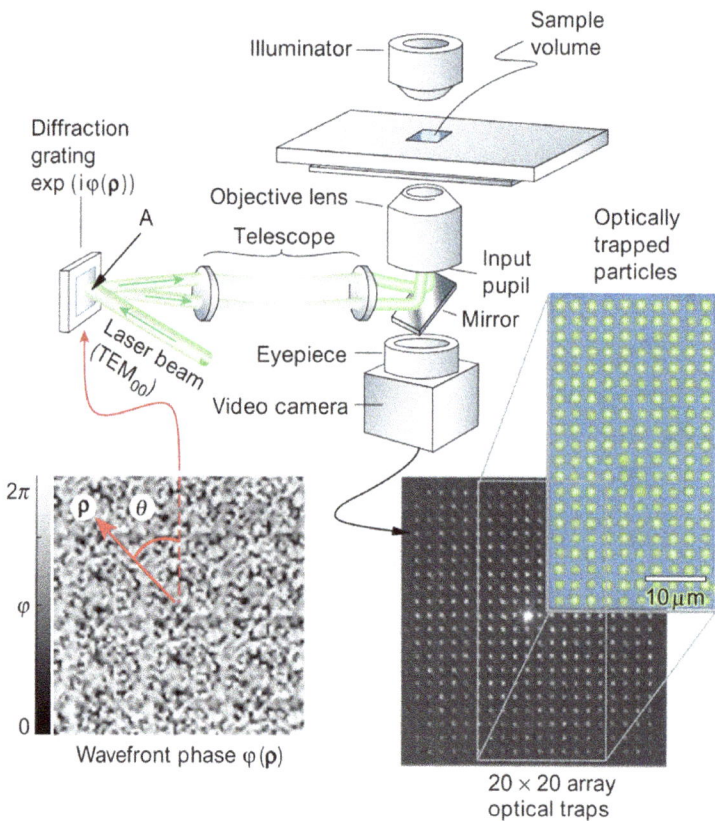

Figure 7.4. Creation of a large number of optical traps by using computer-generated holograms. Projecting a collimated Gaussian (TEM$_{00}$) laser beam through the input pupil of a strongly converging lens, such as a microscope objective, creates a single optical trap. The telescope creates an image of the objective's input pupil, centred at point A. Multiple beams passing through point A pass into the objective lens to create multiple optical traps. A single Gaussian laser beam can be spilt into an arbitrary fan-out of beams, all emanating from point A, by an appropriate computer-designed diffraction grating. The example phase grating $\varphi(\rho)$ creates the 20 × 20 array of traps shown in the micrograph. These are shown trapping 800 nm diameter polystyrene spheres dispersed in water. Reproduced by permission from reference [52].

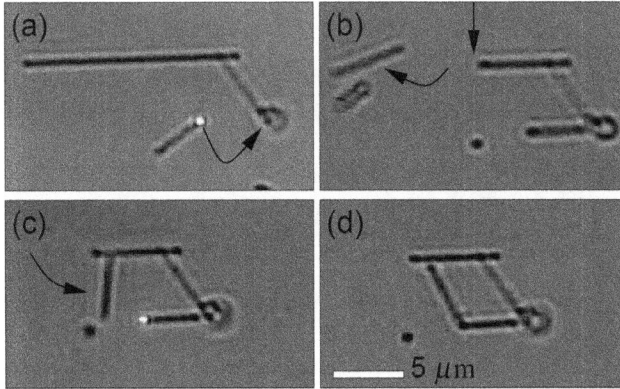

Figure 7.5. Assembly of a rhombus from semiconductor nanowires using holographic optical traps. (a) A nanowire is translated towards an existing structure, created earlier by trapping and fusing two nanowires. (b) The long nanowire is cut with a pulsed optical scalpel. (c) The resulting free-floating nanowire piece is brought towards the partially completed structure. (d) The structure is completed by fusing both ends of the nanowire piece. Reproduced by permission from reference [66].

holographic optical traps. Although optical lattices are usually created with Gaussian beams, systems based on Laguerre-Gaussian [52, 69] or Bessel [70] laser light have been demonstrated, which produce more highly structured forms of trap. Complex beams such as these are discussed in detail in the following chapter.

References

[1] Burns M M, Fournier J-M and Golovchenko J A 1990 Optical matter: crystallization and binding in intense optical fields *Science* **249** 749–54

[2] Chiou A E, Wang W, Sonek G J, Hong J and Berns M W 1997 Interferometric optical tweezers *Opt. Commun.* **133** 7–10

[3] Zemánek P, Jonáš A, Šrámek L and Liška M 1998 Optical trapping of Rayleigh particles using a Gaussian standing wave *Opt. Commun.* **151** 273–85

[4] Bloch I 2005 Ultracold quantum gases in optical lattices *Nat. Phys.* **1** 23–30

[5] Li T C, Kelkar H, Medellin D and Raizen M G 2008 Real-time control of the periodicity of a standing wave: an optical accordion *Opt. Express* **16** 5465–70

[6] Jessen P S and Deutsch I H 1996 Optical lattices *Advances in Atomic, Molecular, and Optical Physics* ed B Benjamin and W Herbert (Cambridge, MA: Academic Press) pp 95–138

[7] Verkerk P, Lounis B, Salomon C, Cohen-Tannoudji C, Courtois J Y and Grynberg G 1992 Dynamics and spatial order of cold cesium atoms in a periodic optical potential *Phys. Rev. Lett.* **68** 3861–64

[8] Raithel G, Birkl G, Kastberg A, Phillips W D and Rolston S L 1997 Cooling and localization dynamics in optical lattices *Phys. Rev. Lett.* **78** 630–3

[9] Jaksch D, Bruder C, Cirac J I, Gardiner C W and Zoller P 1998 Cold bosonic atoms in optical lattices *Phys. Rev. Lett.* **81** 3108–11

[10] Guidoni L and Verkerk P 1999 Optical lattices: cold atoms ordered by light *J. Opt. B: Quantum Semiclass. Opt.* **1** R23–45

[11] Grynberg G and Robilliard C 2001 Cold atoms in dissipative optical lattices *Phys. Rep.* **355** 335–451

[12] Yukalov V I 2009 Cold bosons in optical lattices *Laser Phys.* **19** 1–110

[13] Krutitsky K V 2016 Ultracold bosons with short-range interaction in regular optical lattices *Phys. Rep.* **607** 1–101

[14] Andrews D L, Naguleswaran S and Stedman G E 1998 Phenomenological damping of nonlinear-optical response tensors *Phys. Rev.* A **57** 4925–9

[15] Moiseyev N, Šindelka M and Cederbaum L S 2007 Trapping of cold atoms in optical lattices by the quadrupole force *Phys. Lett.* A **362** 215–20

[16] Anderson B P and Kasevich M A 1998 Macroscopic quantum interference from atomic tunnel arrays *Science* **282** 1686–9

[17] Ben Dahan M, Peik E, Reichel J, Castin Y and Salomon C 1996 Bloch oscillations of atoms in an optical potential *Phys. Rev. Lett.* **76** 4508–11

[18] Kolovsky A R and Korsch H J 2003 Bloch oscillations of cold atoms in two-dimensional optical lattices *Phys. Rev.* A **67** 063601

[19] Lewenstein M, Sanpera A, Ahufinger V, Damski B, Sen A and Sen U 2007 Ultracold atomic gases in optical lattices: mimicking condensed matter physics and beyond *Adv. Phys.* **56** 243–379

[20] Deutsch I H, Brennen G K and Jessen P S 2000 Quantum computing with neutral atoms in an optical lattice *Fortschr. Phys.* **48** 925–43

[21] Monroe C 2002 Quantum information processing with atoms and photons *Nature* **416** 238–46

[22] Pachos J K and Knight P L 2003 Quantum computation with a one-dimensional optical lattice *Phys. Rev. Lett.* **91** 107902

[23] Porto J V, Rolston S, Laburthe Tolra B, Williams C J and Phillips W D 2003 Quantum information with neutral atoms as qubits *Philos. Trans. R. Soc.* A **361** 1417–27

[24] Jaksch D 2004 Optical lattices, ultracold atoms and quantum information processing *Contemp. Phys.* **45** 367–81

[25] Treutlein P *et al* 2006 Quantum information processing in optical lattices and magnetic microtraps *Fortschr. Phys.* **54** 702–18

[26] Bloch I, Dalibard J and Nascimbene S 2012 Quantum simulations with ultracold quantum gases *Nat. Phys.* **8** 267–76

[27] Amico L, Fazio R, Osterloh A and Vedral V 2008 Entanglement in many-body systems *Rev. Mod. Phys.* **80** 517–76

[28] Brennen G K, Caves C M, Jessen P S and Deutsch I H 1999 Quantum logic gates in optical lattices *Phys. Rev. Lett.* **82** 1060–3

[29] Brennen G K, Deutsch I H and Jessen P S 2000 Entangling dipole-dipole interactions for quantum logic with neutral atoms *Phys. Rev.* A **61** 062309

[30] Jaksch D, Briegel H J, Cirac J I, Gardiner C W and Zoller P 1999 Entanglement of atoms via cold controlled collisions *Phys. Rev. Lett.* **82** 1975–8

[31] Briegel H J, Calarco T, Jaksch D, Cirac J I and Zoller P 2000 Quantum computing with neutral atoms *J. Mod. Opt.* **47** 415–51

[32] Calarco T, Dorner U, Julienne P S, Williams C J and Zoller P 2004 Quantum computations with atoms in optical lattices: Marker qubits and molecular interactions *Phys. Rev.* A **70** 012306

[33] Weitenberg C, Endres M, Sherson J F, Cheneau M, Schausz P, Fukuhara T, Bloch I and Kuhr S 2011 Single-spin addressing in an atomic Mott insulator *Nature* **471** 319–24

[34] Takamoto M, Hong F-L, Higashi R and Katori H 2005 An optical lattice clock *Nature* **435** 321-4

[35] Bloom B J, Nicholson T L, Williams J R, Campbell S L, Bishof M, Zhang X, Zhang W, Bromley S L and Ye J 2014 An optical lattice clock with accuracy and stability at the 10^{-18} level *Nature* **506** 71-5

[36] Ludlow A D, Boyd M M, Ye J, Peik E and Schmidt P O 2015 Optical atomic clocks *Rev. Mod. Phys.* **87** 637-701

[37] Derevianko A and Katori H 2011 Physics of optical lattice clocks *Rev. Mod. Phys.* **83** 331-47

[38] Le Targat R *et al* 2013 Experimental realization of an optical second with strontium lattice clocks *Nat. Commun.* **4** 2109

[39] Sasaki K, Koshioka M, Misawa H, Kitamura N and Masuhara H 1991 Pattern formation and flow control of fine particles by laser-scanning micromanipulation *Opt. Lett.* **16** 1463-5

[40] Sasaki K, Koshioka M, Misawa H, Kitamura N and Masuhara H 1992 Optical trapping of a metal particle and a water droplet by a scanning laser beam *Appl. Phys. Lett.* **60** 807-9

[41] Mio C, Gong T, Terray A and Marr D W M 2000 Design of a scanning laser optical trap for multiparticle manipulation *Rev. Sci. Instrum.* **71** 2196-200

[42] Visscher K, Gross S P and Block S M 1996 Construction of multiple-beam optical traps with nanometer-resolution position sensing *IEEE J. Sel. Top. Quant. Electron* **2** 1066-76

[43] Vossen D L J, van der Horst A, Dogterom M and van Blaaderen A 2004 Optical tweezers and confocal microscopy for simultaneous three-dimensional manipulation and imaging in concentrated colloidal dispersions *Rev. Sci. Instrum.* **75** 2960-70

[44] Valentine M T, Guydosh N R, Gutiérrez-Medina B, Fehr A N, Andreasson J O and Block S M 2008 Precision steering of an optical trap by electro-optic deflection *Opt. Lett.* **33** 599-601

[45] Čižmár T, Dávila Romero L C, Dholakia K and Andrews D L 2010 Multiple optical trapping and binding: New routes to self-assembly *J. Phys. B: At. Mol. Opt. Phys.* **43** 102001

[46] Grier D G 2003 A revolution in optical manipulation *Nature* **424** 810-6

[47] Woerdemann M, Alpmann C, Esseling M and Denz C 2013 Advanced optical trapping by complex beam shaping *Laser & Photon. Rev* **7** 839-54

[48] Dufresne E R and Grier D G 1998 Optical tweezer arrays and optical substrates created with diffractive optics *Rev. Sci. Instrum.* **69** 1974-7

[49] Dufresne E R, Spalding G C, Dearing M T, Sheets S A and Grier D G 2001 Computer-generated holographic optical tweezer arrays *Rev. Sci. Instrum.* **72** 1810-6

[50] Reicherter M, Haist T, Wagemann E U and Tiziani H J 1999 Optical particle trapping with computer-generated holograms written on a liquid-crystal display *Opt. Lett.* **24** 608-10

[51] Liesener J, Reicherter M, Haist T and Tiziani H J 2000 Multi-functional optical tweezers using computer-generated holograms *Opt. Commun.* **185** 77-82

[52] Curtis J E, Koss B A and Grier D G 2002 Dynamic holographic optical tweezers *Opt. Commun.* **207** 169-75

[53] Hossack W J, Theofanidou E, Crain J, Heggarty K and Birch M 2003 High-speed holographic optical tweezers using a ferroelectric liquid crystal microdisplay *Opt. Express* **11** 2053-9

[54] Melville H, Milne G F, Spalding G C, Sibbett W, Dholakia K and McGloin D 2003 Optical trapping of three-dimensional structures using dynamic holograms *Opt. Express* **11** 3562-7

[55] Leach J, Sinclair G, Jordan P, Courtial J, Padgett M J, Cooper J and Laczik Z J 2004 3D manipulation of particles into crystal structures using holographic optical tweezers *Opt. Express* **12** 220-6

[56] Sinclair G, Leach J, Jordan P, Gibson G, Yao E, Laczik Z J, Padgett M J and Courtial J 2004 Interactive application in holographic optical tweezers of a multi-plane Gerchberg-Saxton algorithm for three-dimensional light shaping *Opt. Express* **12** 1665–70

[57] Sinclair G, Jordan P, Leach J, Padgett M J and Cooper J 2004 Defining the trapping limits of holographical optical tweezers *J. Mod. Opt.* **51** 409–14

[58] Schmitz C H J, Spatz J P and Curtis J E 2005 High-precision steering of multiple holographic optical traps *Opt. Express* **13** 8678–85

[59] Polin M, Ladavac K, Lee S-H, Roichman Y and Grier D G 2005 Optimized holographic optical traps *Opt. Express* **13** 5831–45

[60] Grier D G and Roichman Y 2006 Holographic optical trapping *Appl. Opt.* **45** 880–7

[61] Mogensen P C and Glückstad J 2000 Dynamic away generation and pattern formation for optical tweezers *Opt. Commun.* **175** 75–81

[62] Eriksen R L, Daria V R and Glückstad J 2002 Fully dynamic multiple-beam optical tweezers *Opt. Express* **10** 597–602

[63] Rodrigo P J, Perch-Nielsen I R, Alonzo C A and Glückstad J 2006 GPC-based optical micromanipulation in 3D real-time using a single spatial light modulator *Opt. Express* **14** 13107–12

[64] Dam J S, Rodrigo P J, Perch-Nielsen I R, Alonzo C A and Glückstad J 2007 Computerized 'drag-and-drop' alignment of GPC-based optical micromanipulation system *Opt. Express* **15** 1923–31

[65] Plewa J, Tanner E, Mueth D M and Grier D G 2004 Processing carbon nanotubes with holographic optical tweezers *Opt. Express* **12** 1978–81

[66] Agarwal R, Ladavac K, Roichman Y, Yu G, Lieber C M and Grier D G 2005 Manipulation and assembly of nanowires with holographic optical traps *Opt. Express* **13** 8906–12

[67] Simpson S H and Hanna S 2010 Holographic optical trapping of microrods and nanowires *J. Opt. Soc. Am.* A **27** 1255–64

[68] Li J and Du G 2014 Manipulation and assembly of ZnO nanowires with single holographic optical tweezers system *Appl. Opt.* **53** 351–5

[69] Curtis J E and Grier D G 2003 Structure of optical vortices *Phys. Rev. Lett.* **90** 133901

[70] Čižmár T, Kollárová V, Tsampoula X, Gunn-Moore F, Sibbett W, Bouchal Z and Dholakia K 2008 Generation of multiple Bessel beams for a biophotonics workstation *Opt. Express* **16** 14024–35

Chapter 8

Orbital angular momentum, optical vortices and torques

As we observed in chapter 2, it is possible for beams of light to convey more than simply energy, linear momentum and spin angular momentum. When suitably structured, light can also deliver orbital angular momentum (OAM), whose quantised form is designated by eigenstates and eigenvalues of the operator $\hat{L}_{rad}$. Any such eigenstate is commonly referred to as an *optical vortex*, alternatively called *twisted light*. Achieving this type of structure places constraints upon the form of the transverse field distribution, specific forms of which are considered below. However, the key attribute is, in each case, a phase factor of the form $\exp(\pm il\phi)$, where l denotes a *topological charge* or *winding number* and ϕ is the azimuthal angle around the beam axis. Generally, l signifies the number of distinct helical surfaces formed by the optical wavefront, as it winds around this axis within the span of a wavelength— see figure 8.1. Both the electric and magnetic fields exhibit this feature. One immediate consequence is that, for $l \neq 0$, there has to be a line of zero intensity along the axis, as a result of an indeterminate phase at the corresponding point in the transverse plane. This feature lends its character to another name associated with this type of beam: *singular optics*. Light engendered with OAM includes *Laguerre-Gaussian*, and all but the simplest forms of *Bessel* and *Mathieu* beams [1–3]. Within the high-intensity rings of such structured light, particles may be trapped and *rotated* (as shown by figure 8.2); the presence of an optical torque due to the beam 'twist' is the origin of the latter. Outside of optical manipulation, a potentially important industrial application of structured light is quantum informatics. This prospect arises due to the additional information offered by each photon of the complex beam, beyond that afforded by the simple binary basis of polarisation [4].

doi:10.1088/978-1-6817-4465-0ch8

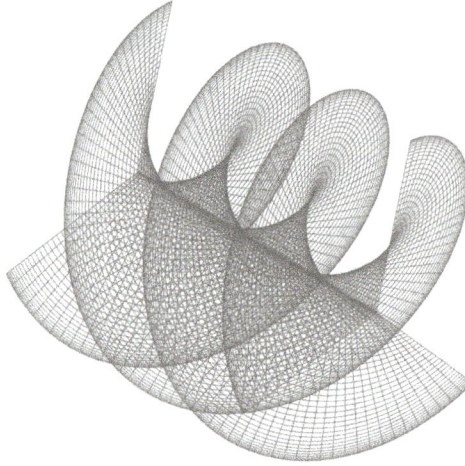

Figure 8.1. Wave-front surface for an optical vortex with a topological charge of three, over the span of three wavelengths (three intertwined helical surfaces).

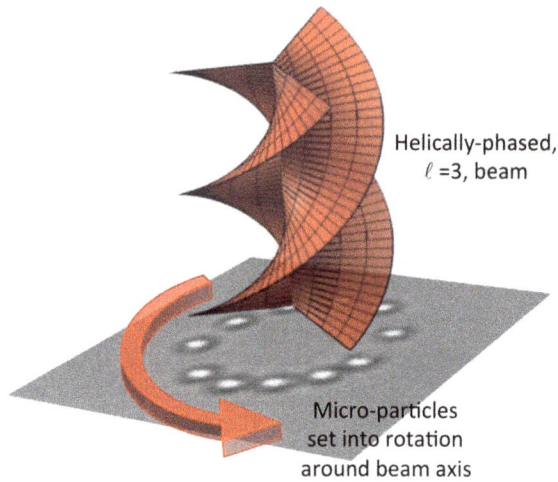

Helically-phased,
ℓ =3, beam

Micro-particles
set into rotation
around beam axis

Figure 8.2. Helically phased laser beam incident on a set of microscopic particles. The latter are trapped within the high-intensity ring of the beam (the centre of an optical vortex beam has zero intensity, unlike Gaussian light) and orbit around the beam axis, as indicated by the red arrow. Image courtesy of Miles Padgett, University of Glasgow.

Orbital angular momentum

Experimental methods for bestowing a conventional Gaussian beam with OAM include passage of the beam through pitch-fork holograms [5–7], spiral phase plates [8], paired cylindrical lenses [9], q-plates [10], hyperbolic metamaterials [11] or SLMs [12–14]. Moreover, a theoretical method has been proposed for the direct generation of light endowed with OAM, based on light emission from an array of nanoparticles [15–17]. To understand the potential that the light then acquires for manipulating nanoscale material, we now return to the theory of optical angular momentum, introduced in chapter 2.

The total angular momentum can be written as $\hat{\mathbf{J}}_{rad} = \hat{\mathbf{S}}_{rad} + \hat{\mathbf{L}}_{rad}$ (although this separation is only possible by use of the paraxial approximation; non-paraxial light is not considered in this book, but has been examined elsewhere [18–21]). Here, the spin angular momentum, $\hat{\mathbf{S}}_{rad}$, is given by equation (2.12) and the OAM operator, $\hat{\mathbf{L}}_{rad}$, is expressible as;

$$\hat{\mathbf{L}}_{rad} = \sum_{k,\eta,l,p} \hat{N}_{lp}^{(\eta)}(\mathbf{k}) l\hbar \mathbf{u_k}, \tag{8.1}$$

which is cast in terms of Laguerre-Gaussian modes with four degrees of freedom: a wave-vector of magnitude k, polarisation η, and integers l and p that designate the order and degree of the corresponding Laguerre polynomial[1]. The latter symbol is the radial index that represents the number of radial nodes, $p + 1$, within the transverse field distribution of the beam. In passing, it is interesting to note that by casting the sum over η explicitly in terms of a basis comprising left- and right-handed circular polarisations, we arrive at [22];

$$\hat{\mathbf{L}}_{rad} = \sum_{k,l,p} \left\{ \hat{N}_{lp}^{(L)}(\mathbf{k}) + \hat{N}_{lp}^{(R)}(\mathbf{k}) \right\} l\hbar \mathbf{u_k}, \tag{8.2}$$

which forms a neat counterpart to the spin angular momentum given by equation (2.12). The above expressions indicate that the orbital angular momentum has no connection with any spin helicity associated with polarisation [23]. Furthermore, we can determine that a free-field photon endowed with topological charge has an OAM magnitude of $l\hbar$.

Optical vortices

The theoretical basis for the concept of an optical vortex beam was first established in a series of works [1, 24, 25]. For Laguerre-Gaussian modes, the explicit formulations for the electric and magnetic field expansions—previously defined in equations (2.3) and (2.4) for unstructured light—are written as follows [26];

$$\hat{\mathbf{E}}(\mathbf{r}) = i \sum_{k,\eta,l,p} \left(\frac{\hbar ck}{\varepsilon_0 V}\right)^{\frac{1}{2}} f_{l,p}(r) \left\{ \mathbf{e}_{l,p}^{(\eta)}(\mathbf{k}) \hat{a}_{l,p}^{(\eta)}(\mathbf{k}) e^{i(\mathbf{k}\cdot\mathbf{r}+l\phi)} - \bar{\mathbf{e}}_{l,p}^{(\eta)}(\mathbf{k}) \hat{a}_{l,p}^{\dagger(\eta)}(\mathbf{k}) e^{-i(\mathbf{k}\cdot\mathbf{r}+l\phi)} \right\}, \tag{8.3}$$

$$\hat{\mathbf{B}}(\mathbf{r}) = i \sum_{k,\eta,l,p} \left(\frac{\hbar k}{\varepsilon_0 c V}\right)^{\frac{1}{2}} f_{l,p}(r) \left\{ \mathbf{b}_{l,p}^{(\eta)}(\mathbf{k}) \hat{a}_{l,p}^{(\eta)}(\mathbf{k}) e^{i(\mathbf{k}\cdot\mathbf{r}+l\phi)} - \bar{\mathbf{b}}_{l,p}^{(\eta)}(\mathbf{k}) \hat{a}_{l,p}^{\dagger(\eta)}(\mathbf{k}) e^{-i(\mathbf{k}\cdot\mathbf{r}+l\phi)} \right\}, \tag{8.4}$$

using cylindrical coordinates that comprise off-axial radial distance r, axial position z and azimuthal angle ϕ. In these expressions, $f_{l,p}(r)$ is a normalised radial function,

---

[1] In comparison a Gaussian beam in a specified direction has two degrees of freedom, k and η.

whose form depends on *l* only through its modulus. In consequence, all information on wavefront *handedness* is found within the phase factors, in which a positive or negative sign for *l* represents a helical wavefront structure of left- or right-handed helicity, respectively. Typical examples of transverse field distributions for Laguerre-Gaussian modes are presented in figure 8.3, which specifically depicts the associated intensity and phase structure. Figure 8.4(a) gives an impression of

Figure 8.3. Transverse field distributions for five Laguerre-Gaussian modes, with their associated intensity and phase structure. Top images display $l = 1$ (left) and $l = -1$ (right) with $p = 0$, while the lower images show $l = 3$ with $p = 0, 1, 2$ (left, centre, right). White areas represent zero intensity and the hues denote the phase.

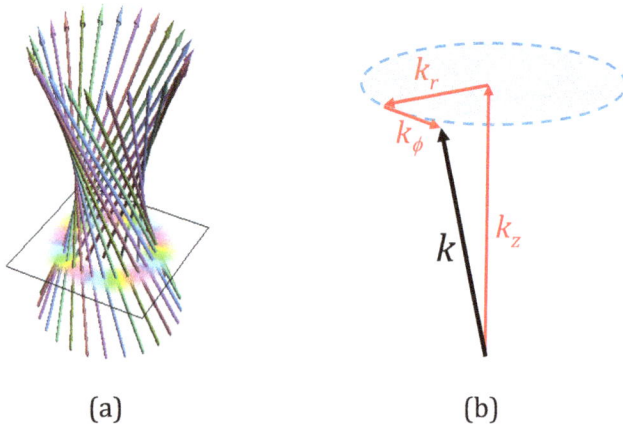

(a)   (b)

Figure 8.4. (a) Schematic representation for the focus of an $l = 5$ optical vortex mode. In the transverse plane, hue indicates the phase and brightness the intensity. The bundle of Poynting vectors, depicted as colour-coded arrows, rotates about the beam axis by one fifth of a circle over the space of one wavelength: (b) resolution of a local Poynting vector into axial (z), radial (r) and azimuthal (ϕ) components. Image (a) courtesy of Sonja Franke-Arnold, University of Glasgow.

how the wave-vector, everywhere pointing along the normal to the wavefront surface, effectively twists around the beam axis. Laguerre-Gaussian beams with $p > 0$ are, in a sense, similar in nature to higher-order ($l > 0$) Bessel beams; however, the latter has non-diffracting and self-healing properties [27–30]. The electric and magnetic field expansions for the Bessel beam are identical to equations (8.3) and (8.4), except that $f_{l,p}(r)$ is substituted with an lth-order Bessel function. Mathieu beams are essentially Bessel beams with elliptical (rather than circular) rings, with field expressions correspondingly similar in form [31, 32].

Optical torque

Beams of light with various structures (both those with and without OAM) can trap and manipulate nanoparticles. The particles are typically attracted to the high intensity regions via the gradient force – i.e. the circular rings in the case of a Laguerre-Gaussian or a higher-order Bessel beam, as opposed to the centre of a regular Gaussian beam. However, as an additional feature, beams that contain OAM may also produce an optical torque. It should be emphasised that such a torque is not the same as the one expressed in equation (3.9), which corresponds to the alignment of a non-spherical nanoparticle with the polarisation plane of the trapping beam due to the gradient force. The optical torque of present interest occurs even with spherical particles, producing particle circulation as indicated by the arrow within figure 8.2. This torque may arise due to the radiation force associated with *non-forward* Rayleigh scattering, rather than the force detailed in chapter 3 that relates to light absorption. The radiation force will usually 'push' the particles along the beam axis (explaining the need for a beam focus in the optical tweezers set-up, as stated in chapter 5). However, the inclusion of a helical wavefront means that a torque will also be applied to the nanoparticle. A quantum description of such a torque now follows.

The optical torque that originates from the helical wavefront can be determined from second-order perturbation theory. Here, in contrast to equation (3.6), the initial and final system states are *not* identical since the torque is based on a non-forward Rayleigh scattering mechanism. In such a process, the created and annihilated photons have identical energy but travel in different directions. Due to this disparity an additional radiation mode is assigned to the system states, so that $|I\rangle = |n, 0; 0\rangle$ and $|F\rangle = |(n − 1), 1; 0\rangle$ for the present case. Here, the first and second elements in each state vector denote radiation modes relating to the annihilation and creation events, respectively, and the third element represents the particle state. (Forward Rayleigh scattering, in contrast, entails photon annihilation and creation from and into the same radiation mode.) This subtle change in the initial and final system states, which contrasts to the forward Rayleigh scattering case of chapter 3, may seem innocuous but it facilitates a physically important differentiation between the radiation and gradient force. The optical torque is determined by multiplying the scattering rate (found from the Fermi rule) by the radius of the beam ring, r_b, and the orbital angular momentum per photon $l\hbar\mathbf{u}_{kz}$, where $\mathbf{u}_{kz}$ is the unit vector of $\mathbf{k}_z$; the latter is illustrated in figure 8.4(b). Following a rotational-average, assuming plane polarised input and summing over all possible polarisations for the scattered light, the following expression is derived for the observed torque;

$$\tau = \frac{Ik^3 lr_b \mathbf{u_{kz}}}{40\pi\varepsilon_0^2 c}\left(\alpha_{\lambda\lambda}\alpha_{\mu\mu} + 7\alpha_{\lambda\mu}\alpha_{\lambda\mu}\right). \tag{8.5}$$

Here, the Greek indices denote the particle frame of reference for the polarisability components (rather than the laboratory frame represented by the earlier Latin indices).

In equation (8.5), the optical response of the particle is written in terms of two material parameters, each written using the implied summation convention encountered in chapter 3. Thus, $\alpha_{\lambda\lambda}\alpha_{\mu\mu}$ represents the square of the trace of the polarisability, $\text{tr}(\boldsymbol{\alpha})$, and $\alpha_{\lambda\mu}\alpha_{\lambda\mu}$ is a sum of the squares of the nine polarisability components. Larger dielectric particles may be treated with electronic properties closer to those of a bulk material. It is appropriate, in such cases, to engage the linear optical susceptibility rather than the polarisability tensor; the connection between the two is explained in detail elsewhere [33]. In all situations, therefore, there are just two scalar parameters, which together effectively determine the particle's propensity to acquire angular momentum about the vortex beam axis.

The circular motion produced by action of an optical vortex will, of course, be subject to viscous forces acting as a drag, and these are simply determined by the standard viscosity of the host medium. The velocity is thus given by $v = \tau/c_d r_b$, where c_d is the drag coefficient. This orbiting behaviour was first theoretically predicted [34] and experimentally observed [35–39] with Laguerre-Gaussian light (sometimes known as an optical spanner or wrench)—such a system is now recognised as a foundation for the next generation of optical tweezers [40–43]. The technique has since been refined by a scheme involving the interference pattern of a Laguerre-Gaussian beam with a Gaussian laser, which circumvents problems such as the photodamage of particles due to heating. In such a scheme the number of 'arms' of the spiral pattern, produced by the beam interference and relating to the l value of the structured beam, can be tailored to fit the shape of the particle. Moreover, by changing the effective pathlength of one of the interfering beams, the speed of the orbit can be optically controlled [44]. In some of the most exotic structures, the circulation of particles within interlocking vortex rings (figure 8.5) and, elsewhere, a higher-order Bessel beam [45] have also been demonstrated.

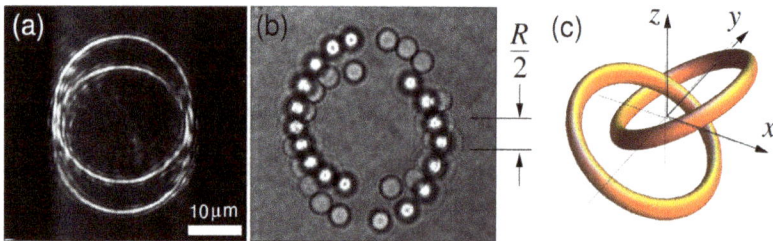

Figure 8.5. (a) Interlocked optical vortex traps; (b) particles circulating about each trap (note the depth profile); (c) depiction of the three-dimensional configuration. Image courtesy of David Grier, New York University.

References

[1] Allen L, Beijersbergen M W, Spreeuw R J C and Woerdman J P 1992 Orbital angular momentum of light and the transformation of Laguerre-Gaussian laser modes *Phys. Rev. A* **45** 8185–89

[2] Chávez-Cerda S, Padgett M J, Allison I, New G H C, Gutiérrez-Vega J C, O'Neil A T, MacVicar I and Courtial J 2002 Holographic generation and orbital angular momentum of high-order mathieu beams *J. Opt. B: Quantum Semiclass. Opt.* **4** S52–57

[3] Volke-Sepulveda K, Garcés-Chávez V, Chávez-Cerda S, Arlt J and Dholakia K 2002 Orbital angular momentum of a high-order bessel light beam *J. Opt. B: Quantum Semiclass. Opt.* **4** S82–89

[4] Xie G *et al* 2015 Performance metrics and design considerations for a free-space optical orbital-angular-momentum-multiplexed communication link *Optica* **2** 357–65

[5] Bazhenov V Y, Vasnetsov M and Soskin M 1990 Laser beams with screw dislocations in their wavefronts *JETP Lett.* **52** 429–31

[6] Bazhenov V Y, Soskin M and Vasnetsov M 1992 Screw dislocations in light wavefronts *J. Mod. Opt.* **39** 985–90

[7] Mirhosseini M, Magana-Loaiza O S, Chen C, Rodenburg B, Malik M and Boyd R W 2013 Rapid generation of light beams carrying orbital angular momentum *Opt. Express* **21** 30196–203

[8] Beijersbergen M W, Coerwinkel R P C, Kristensen M and Woerdman J P 1994 Helical-wave-front laser-beams produced with a spiral phase plate *Opt. Commun.* **112** 321–27

[9] Beijersbergen M W, Allen L, Vanderveen H E L O and Woerdman J P 1993 Astigmatic laser mode converters and transfer of orbital angular-momentum *Opt. Commun.* **96** 123–32

[10] Marrucci L, Manzo C and Paparo D 2006 Optical spin-to-orbital angular momentum conversion in inhomogeneous anisotropic media *Phys. Rev. Lett.* **96** 163905

[11] Sun J, Zeng J and Litchinitser N M 2013 Twisting light with hyperbolic metamaterials *Opt. Express* **21** 14975–81

[12] Heckenberg N R, McDuff R, Smith C P and White A G 1992 Generation of optical phase singularities by computer-generated holograms *Opt. Lett.* **17** 221–3

[13] Ostrovsky A S, Rickenstorff-Parrao C and Arrizón V 2013 Generation of the 'perfect' optical vortex using a liquid-crystal spatial light modulator *Opt. Lett.* **38** 534–6

[14] Forbes A, Dudley A and McLaren M 2016 Creation and detection of optical modes with spatial light modulators *Adv. Opt. Photon* **8** 200–27

[15] Coles M M, Williams M D, Saadi K, Bradshaw D S and Andrews D L 2013 Chiral nanoemitter array: a launchpad for optical vortices *Laser Photon. Rev.* **7** 1088–92

[16] Williams M D, Coles M M, Saadi K, Bradshaw D S and Andrews D L 2013 Optical vortex generation from molecular chromophore arrays *Phys. Rev. Lett.* **111** 153603

[17] Williams M D, Coles M M, Bradshaw D S and Andrews D L 2014 Direct generation of optical vortices *Phys. Rev. A* **89** 033837

[18] Barnett S M and Allen L 1994 Orbital angular momentum and nonparaxial light beams *Opt. Commun.* **110** 670–8

[19] van Enk S J and Nienhuis G 1994 Spin and orbital angular momentum of photons *Europhys. Lett.* **25** 497–501

[20] Santamato E 2004 Photon orbital angular momentum: problems and perspectives *Fortschr. Phys.* **52** 1141–53

[21] Bliokh K Y, Alonso M A, Ostrovskaya E A and Aiello A 2010 Angular momenta and spin–orbit interaction of nonparaxial light in free space *Phys. Rev. A* **82** 063825

[22] Coles M M and Andrews D L 2013 Photonic measures of helicity: optical vortices and circularly polarized reflection *Opt. Lett.* **38** 869–71

[23] Andrews D L, Dávila Romero L C and Babiker M 2004 On optical vortex interactions with chiral matter *Opt. Commun.* **237** 133–9

[24] Nye J F and Berry M V 1974 Dislocations in wave trains *Proc. R. Soc. A* **336** 165–90

[25] Coullet P, Gil L and Rocca F 1989 Optical vortices *Opt. Commun.* **73** 403–8

[26] Dávila Romero L C, Andrews D L and Babiker M 2002 A quantum electrodynamics framework for the nonlinear optics of twisted beams *J. Opt. B: Quantum Semiclass. Opt.* **4** S66–72

[27] Bouchal Z and Olivík M 1995 Non-diffractive vector bessel beams *J. Mod. Opt.* **42** 1555–66

[28] McGloin D and Dholakia K 2005 Bessel beams: diffraction in a new light *Contemp. Phys.* **46** 15–28

[29] Mazilu M, Stevenson D J, Gunn-Moore F and Dholakia K 2010 Light beats the spread: 'non-diffracting' beams *Laser Photon. Rev.* **4** 529–47

[30] Aiello A and Agarwal G S 2014 Wave-optics description of self-healing mechanism in Bessel beams *Opt. Lett.* **39** 6819–22

[31] Lóxpez-Mariscal C, Gutiérrez-Vega J C, Milne G and Dholakia K 2006 Orbital angular momentum transfer in helical mathieu beams *Opt. Express* **14** 4182–87

[32] Alpmann C, Bowman R, Woerdemann M, Padgett M and Denz C 2010 Mathieu beams as versatile light moulds for 3D micro particle assemblies *Opt. Express* **18** 26084–91

[33] Andrews D L and Allcock P 2002 *Optical Harmonics in Molecular Systems* (Weinheim: Wiley-VCH) p 10

[34] Babiker M, Power W L and Allen L 1994 Light-induced torque on moving atoms *Phys. Rev. Lett.* **73** 1239–42

[35] He H, Friese M E J, Heckenberg N R and Rubinsztein-Dunlop H 1995 Direct observation of transfer of angular-momentum to absorptive particles from a laser-beam with a phase singularity *Phys. Rev. Lett.* **75** 826–9

[36] Friese M E J, Enger J, Rubinsztein-Dunlop H and Heckenberg N R 1996 Optical angular-momentum transfer to trapped absorbing particles *Phys. Rev. A* **54** 1593–6

[37] Simpson N B, Allen L and Padgett M J 1996 Optical tweezers and optical spanners with Laguerre-Gaussian modes *J. Mod. Opt.* **43** 2485–91

[38] Simpson N B, Dholakia K, Allen L and Padgett M J 1997 Mechanical equivalence of spin and orbital angular momentum of light: an optical spanner *Opt. Lett.* **22** 52–4

[39] Roichman Y, Grier D G and Zaslavsky G 2007 Anomalous collective dynamics in optically driven colloidal rings *Phys. Rev. E* **75** 020401

[40] Molloy J E and Padgett M J 2002 Lights, action: optical tweezers *Contemp. Phys.* **43** 241–58

[41] Dholakia K, MacDonald M P and Spalding G C 2002 Optical tweezers: the next generation *Phys. World* **15** 31–5

[42] Padgett M J and Bowman R 2011 Tweezers with a twist *Nat. Photonics* **5** 343–8

[43] Padgett M J 2014 Light's twist *Proc. R. Soc. A* **470** 20140633

[44] Paterson L, MacDonald M P, Arlt J, Sibbett W, Bryant P E and Dholakia K 2001 Controlled rotation of optically trapped microscopic particles *Science* **292** 912–4

[45] Khonina S N, Kotlyar V V, Skidanov R V, Soifer V A, Jefimovs K, Simonen J and Turunen J 2004 Rotation of microparticles with Bessel beams generated by diffractive elements *J. Mod. Opt.* **51** 2167–84

Chapter 9

Structured light: particle steering, traction and optical lift

Initially, optical manipulation with structured light mainly centred upon Laguerre-Gaussian beams. However, new schemes continue to arise such as particle steering with Airy beams and optical traction using solenoid or Bessel beams. All these systems involve microparticle manipulation, although it is expected that down-scaling to nanoscale sizes is achievable. Such techniques, moreover, rely on the structuring of the trapping beam to produce the novel effects [1]. Nonetheless, for microparticles in the ray optics regime, there is increasing interest in achieving sought mechanical effects by sculpting the particle rather than the light field [2, 3]. An example is the optical lift experienced by an aerofoil-shaped refractive object under the influence of a uniform stream of light, as discussed later in this chapter.

Particle steering

Outside of ultracold atom guiding along Laguerre-Gaussian and Bessel beams [4, 5], another intriguing form of particle steering involves Airy beams—accelerating, non-diffracting beams whose paths of highest intensity do not propagate in a straight line [6–9]. Such beams have trajectories that can be engineered, for example, in a parabolic curve in the simplest form. Analogous to Bessel beams, this type of laser light contains self-healing properties as conveyed by figure 9.1. A particle within an Airy beam is shown to undergo ballistic dynamics, similar to projectiles acting under gravity [10]. Based on this concept a 'snowblower' effect has been realised [11], a description of which now follows.

In the 'snowblowing' effect, so-termed by analogy to the propulsion of snow particles upwards and sideways (in an arc due to gravity), a sample of particles are exposed to an Airy beam. In such a system, the sample is typically separated into four compartments and the peak intensity of the asymmetric Airy beam is centrally positioned. The transverse field distribution of such a beam comprises a high

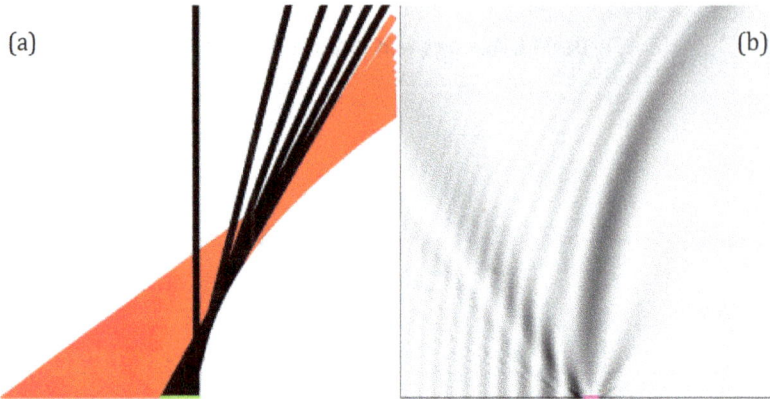

Figure 9.1. Self-healing properties of the Airy beam. (a) Ray tracing picture where black lines represent rays that are blocked by the obstacle (green horizontal line) and the red lines are rays that continue to propagate and reconstruct the Airy beam after some distance. (b) Electromagnetic field representation, in which the greyscale depicts the intensity profile of the reconstructing Airy beam, after an obstacle (pink line) is positioned in front of the first lobe. Adapted from reference [6] with permission from Wiley-VCH Verlag GmBh & Co. KgaA.

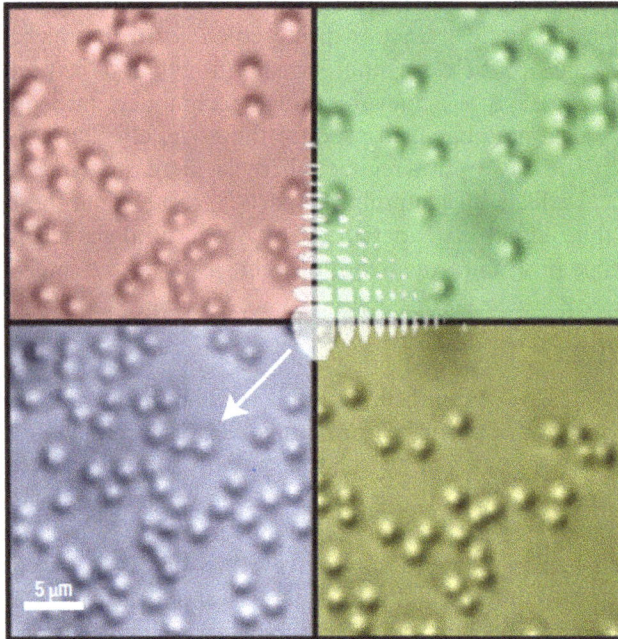

Figure 9.2. Demonstration of the 'snowblower' effect: a sample of colloids transfer from one compartment to another following exposure to an Airy beam for two minutes. The transverse field distribution of the Airy beam (white pattern) and the direction of particle propagation (white arrow) are overlaid; the micrograph is divided into four (coloured) sections. Reproduced by permission from reference [11].

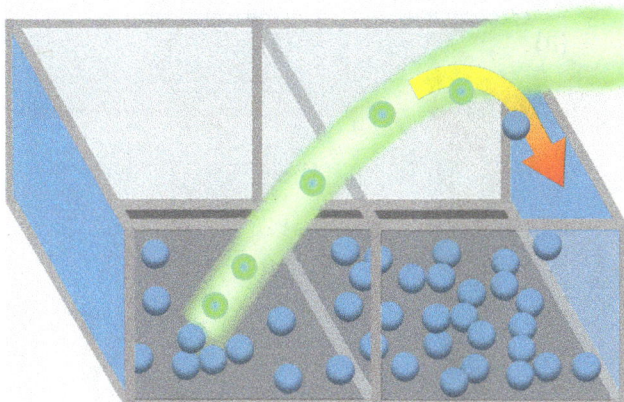

Figure 9.3. Artist's impression of the 'snowblowing' effect. Particles are transferred between compartments because the Airy beam (in green) acts as an optical guide.

intensity, central peak and a number of nearby lobes (with intensities that decay with increasing distance from the centre, as depicted by figure 9.2). The laser light attracts the particles to the high intensity portion of the beam via the gradient force and, subsequently, propels them along a parabolic path due to the radiation force of the curved beam. The steering continues, typically for a distance of around 75 μm, until the beam becomes 'smeared out' and the particles are released. The result, as shown in figures 9.2 and 9.3, is that the Airy beam launches the particles from one compartment to another.

Tractor beams

Typically, as stated in chapter 5, the radiation force of a trapping beam pushes the particles in the same direction as the propagating light. However, in certain circumstances, it has been demonstrated that particles can be transported *towards* the source of the light, in the opposite direction to convention [13]. This phenomenon has been branded a *tractor beam*, a concept that originates from the science fiction genre. A complex form of laser light that may act as an optical tractor is known as a *solenoid beam*, which is in a sense an optical analogue to the Archimedes' screw. Such structured light is characterised by an intensity maximum that spirals round the beam axis—as shown in figure 9.4(a). The upstream movement of a trapped particle, depicted in figure 9.4(b), occurs due to a combination of the gradient and the radiation force of the solenoid beam, i.e. a component of the combined force is directed counter to the light propagation [12, 14][1]. Another form of tractor beam is known as an *optical conveyor*, which can selectively transport illuminated objects either upstream or downstream along the beam axes; such systems can be generated by a superposition of coaxial Bessel beams [15–18]. These conveyors have periodic intensity patterns along their axis (figure 9.5), as a result

---

[1] A Gaussian beam cannot serve in such a capacity since all its momentum is directed along the axis of propagation.

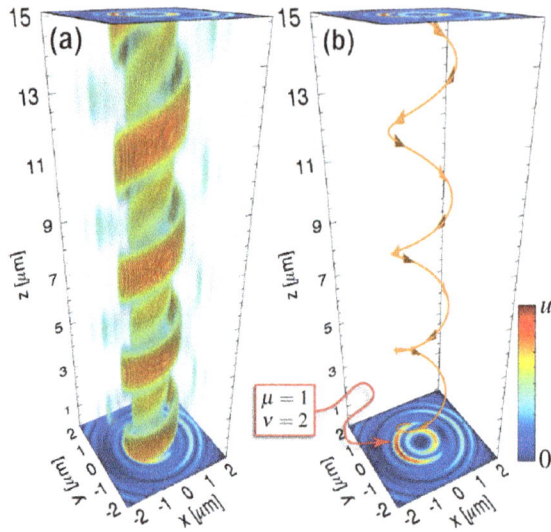

Figure 9.4. (a) Intensity distribution of a right-handed solenoidal tractor beam, with a wavelength of 532 nm, propagating along the +z-direction through water. (b) Trajectory of a 50 nm-diameter silica sphere moving along the tractor beam in the −z-direction. The colour bar indicates the local intensity, u. Reproduced by permission from reference [12]. Copyright 2016 by the American Physical Society.

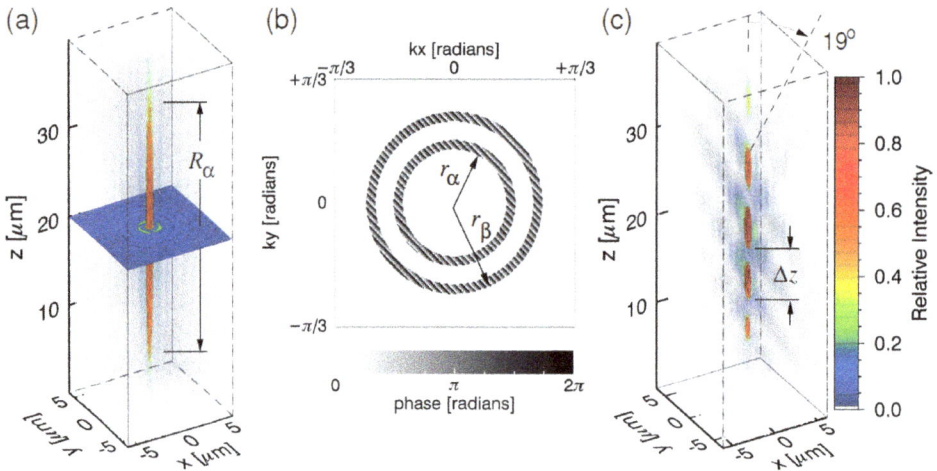

Figure 9.5. (a) Volumetric reconstruction of a Bessel beam projected with a ring-like hologram. (b) Phase hologram encoding an optical conveyor. Diagonal blazing tilts the projected conveyor away from the optical axis. (c) Volumetric reconstruction of the beam projected by the hologram in (b). Adapted from reference [17]. Copyright 2012 by the American Physical Society.

the gradient force may override the push of the radiation force. Varying the relative phase of the beams will shift the axial positions of the optical traps. In this way it is possible to control the passage of the trapped objects downstream or upstream.

Optical lift

Analogous to aerodynamic lift, a cambered refractive object can experience an optical lifting force when placed within a uniform stream of light (figures 9.6 and 9.7). Predicted and observed (in water) by Swartzlander and co-workers, this effect is known as an *optical lift* [19]; the basis for which is refraction and reflection rather than the Bernoulli principle that applies to aerofoils. It has been shown [20] that the optical torque applied to the light-irradiated semi-cylindrical rod, or the 'lightfoil', may vanish at certain angles of attack—enabling a stable

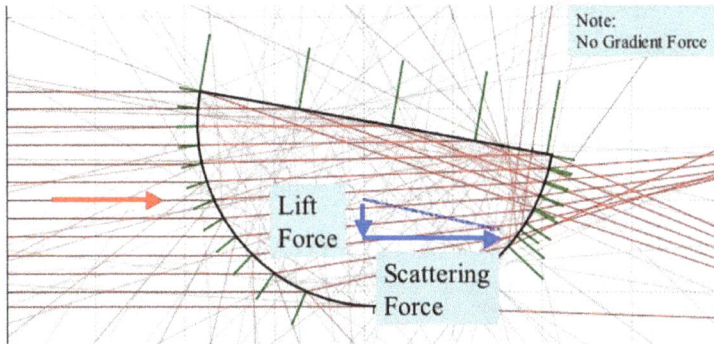

Figure 9.6. Cross-section of a uniformly illuminated semi-cylindrical rod, with the lift force and scattering (radiation) force indicated. Collimated rays (red) are incident from the left and the green lines represent the radiation force for each ray. Image courtesy of Grover Swartzlander, Rochester Institute of Technology.

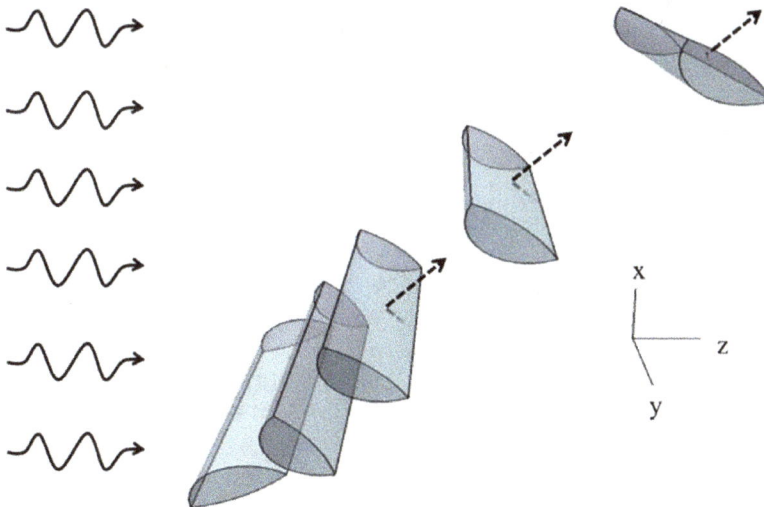

Figure 9.7. Artist's impression of the optical lift of a 'lightfoil' due to laser irradiation. Image courtesy of Grover Swartzlander, Rochester Institute of Technology.

lift. In total, for cases when a light irradiance over 10^3 W cm^{-2} is supplied to overcome gravity, up to four angles of approach are predicted to provide uniform lift without tumbling. This phenomenon originates from the radiation force of the incident light and, unlike optical tweezers, does not involve the gradient force. Due to the latter, numerous rods may be simultaneously lifted in the same radiation field. Many potential applications have been offered, however, the ultimate aim is to deliver space flight systems with navigational and attitude control through solar sails.[21].

References

[1] Dholakia K and Čižmár T 2011 Shaping the future of manipulation *Nat. Photonics* **5** 335–42

[2] Glückstad J 2011 Optical manipulation: sculpting the object *Nat. Photonics* **5** 7–8

[3] Simpson S H, Phillips D B, Carberry D M and Hanna S 2013 Bespoke optical springs and passive force clamps from shaped dielectric particles *J. Quant. Spectrosc. Radiat. Transfer* **126** 91–8

[4] Arlt J, Hitomi T and Dholakia K 2000 Atom guiding along Laguerre-Gaussian and Bessel light beams *Appl. Phys.* B **71** 549–54

[5] Rhodes D P, Gherardi D M, Livesey J, McGloin D, Melville H, Freegarde T and Dholakia K 2006 Atom guiding along high order Laguerre-Gaussian light beams formed by spatial light modulation *J. Mod. Opt.* **53** 547–56

[6] Mazilu M, Stevenson D J, Gunn-Moore F and Dholaktia K 2010 Light beats the spread: 'non-diffracting' beams *Laser Photon. Rev.* **4** 529–47

[7] Siviloglou G A, Broky J, Dogariu A and Christodoulides D N 2007 Observation of accelerating Airy beams *Phys. Rev. Lett.* **99** 213901

[8] Broky J, Siviloglou G A, Dogariu A and Christodoulides D N 2008 Self-healing properties of optical Airy beams *Opt. Express* **16** 12880–91

[9] Dholakia K 2008 Optics: Against the spread of the light *Nature* **451** 413

[10] Siviloglou G A, Broky J, Dogariu A and Christodoulides D N 2008 Ballistic dynamics of Airy beams *Opt. Lett.* **33** 207–9

[11] Baumgartl J, Mazilu M and Dholakia K 2008 Optically mediated particle clearing using Airy wavepackets *Nat. Photonics* **2** 675–8

[12] Yevick A, Ruffner D B and Grier D G 2016 Tractor beams in the Rayleigh limit *Phys. Rev.* A **93** 043807

[13] Dogariu A, Sukhov S and Sáenz J 2013 Optically induced 'negative forces' *Nat. Photonics* **7** 24–7

[14] Lee S-H, Roichman Y and Grier D G 2010 Optical solenoid beams *Opt. Express* **18** 6988–93

[15] Čižmár T, Garcés-Chávez V, Dholakia K and Zemánek P 2005 Optical conveyor belt for delivery of submicron objects *Appl. Phys. Lett.* **86** 174101

[16] Čižmár T, Kollárová V, Bouchal Z and Zemánek P 2006 Sub-micron particle organization by self-imaging of non-diffracting beams *New J. Phys.* **8** 43

[17] Ruffner D B and Grier D G 2012 Optical conveyors: a class of active tractor beams *Phys. Rev. Lett.* **109** 163903

[18] Ruffner D B and Grier D G 2014 Universal, strong and long-ranged trapping by optical conveyors *Opt. Express* **22** 26834–43

[19] Swartzlander G A, Peterson T J, Artusio-Glimpse A B and Raisanen A D 2011 Stable optical lift *Nat. Photonics* **5** 48–51

[20] Simpson S H, Hanna S, Peterson T J and Swartzlander G A 2012 Optical lift from dielectric semicylinders *Opt. Lett.* **37** 4038–40

[21] Artusio-Glimpse A B, Peterson T J and Swartzlander G A 2013 Refractive optical wing oscillators with one reflective surface *Opt. Lett.* **38** 935–7

Chapter 10

Optofluidics: lab-on-a-chip mixing and actuating flow

As we have seen, the optical forces that light exerts on matter have a variety of origins, yet in terms of application they fall naturally into essentially two groups. There are those in which a relatively insignificant resistance to optically applied forces leads to motion that is constrained only, or primarily, by a positional (or angular) variation in the associated force field; optical tweezers is an obvious example. Where such local variation is less pronounced—or is at least monotonic—then sustained motion may ensue, as for example occurs in optical methods of particle separation and cell sorting. All such cases, where the responsible forces are determined only by spatial position, satisfy one of the several alternative criteria for a *dynamically conservative system*.

However there are also systems, especially fluids, in which the aim of achieving optically driven motion is frustrated by other local forces: friction or viscosity are the most typical manifestation. Thus, particles in liquid suspension driven 'downstream' along the axis of a laser beam by radiation pressure attain a terminal velocity, typically of the order 10 μm s^{-1}, as a result of velocity-dependent resistance. Some of the energy content of the driving beam can, therefore, be spent in sustaining constant velocity motion through the host medium. Thermal effects that produce rapidly fluctuating, stochastic variations in local force fields can also modify and constrain uniform motion. Nonetheless, for the system as a whole—that is, the particles of interest, the light and the surrounding medium—energy is, of course, always conserved.

Optical manipulation in microfluidics

Using light to manipulate fluids, or particles within fluid media, is a concept described by the umbrella term *optofluidics*. This idea is especially important for lab-on-a-chip devices, in which laboratory functions can be implemented upon a

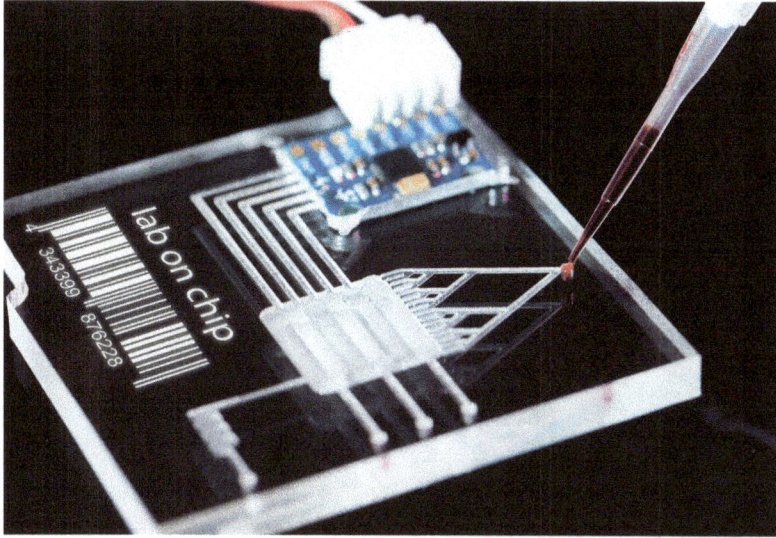

Figure 10.1. Image of a typical lab-on-a-chip device. Copyright Wladimir Bulgar/Shutterstock.com

chip with centimetre- or millimetre-scale dimensions (figure 10.1). Instruments of this type are useful for diagnostics, sensing, testing, pathology and drug discovery. Optical tweezing has an important role in microfluidic systems, such as the mixing of minuscule amounts of chemicals, the actuating of flow and the sorting of micro-particles. (The latter is covered in chapter 6 for cellular bodies; however, the sorting of colloids or similar particles is equally possible in such systems.) It is noteworthy that the practicality of nanofluidics, beyond single molecule sensing, encounters formidable difficulties due to the exceptionally slow volumetric flow of any liquid confined to nanoscale channels [1].

Liquid flow on the microscale generally exhibits laminar (non-turbulent) flow—this arises due to the high resistance against the flow of a typical liquid confined to the microscale; such fluids have a low *Reynolds number*.[1] In application to lab-on-a-chip devices, laminar flow is ideal for transferring minuscule quantities of reagents to a reaction site and the sorting of particles. However, the lack of turbulent flow means that the mixing of the reactants is difficult. This problem can be overcome by a phenomenon introduced in chapter 8: optically trapped (chemically inert) particles that circulate round a ring of a Laguerre-Gaussian beam could induce the turbulence required to instigate the mixing necessary for a chemical reaction. Moreover, amplification of such stirring may be achieved by using a holographic optical tweezers set-up that projects multiple optical vortices [2, 4, 5] (for example, the 3×3 array given by figure 10.2)—interestingly, using the same set of principles, the circulating particles may act as an optomechanical pump to create a flow [6]. Outside of this holographic system, where particles are rotated via OAM transfer, other mixing and pumping schemes based on optical tweezing have been proposed.

---

[1] The Reynolds number is defined as the ratio of inertial forces to viscous forces.

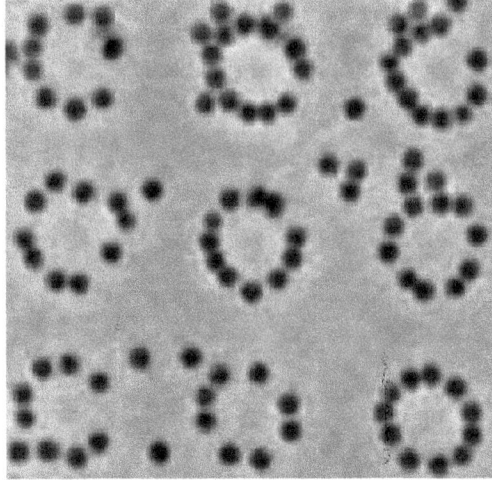

Figure 10.2. Colloidal polystyrene spheres (800 nm in diameter) trapped and circulating within a 3 × 3 array of Laguerre-Gaussian beams, each with $l = 15$. Reproduced by permission from reference [3].

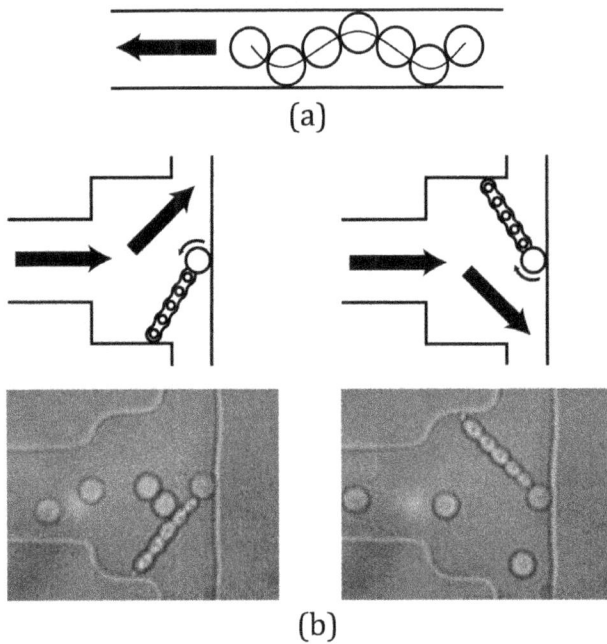

Figure 10.3. (a) Optical pump drives fluid motion through a sinusoidal rippling of a colloidal chain; (b) a valve constructed of colloidal particles in microfluidic channels. Both are activated with optical tweezers, and the direction of fluid flow is indicated by the arrows. The colloidal valve flap is flipped with an optical tweezer and directs particles either upward (left) or downward (right). Actual image of the colloidal peristaltic pump, corresponding to diagram (a), appears in reference [11] and image (b) is reproduced by permission from reference [12].

10-3

Here, momentum is exchanged from a Gaussian beam to a microgear [7, 8], a micropropeller [9] or a calcite crystal [10].

Instead of rotating the microparticles, another optical pump technique involves the translation of particles back and forth in a cooperative manner, as shown by figure 10.3(a). Fabrication of these pumps requires an independent optical trap for each of the microparticles, so that sine wave propagation can be produced to drive the flow of the microscale liquid [11]. An optical valve to actively direct particles into one of two exit channels, as depicted in figure 10.3(b), has also been demonstrated. In this case, the microparticles that constitute the valve structure are rotated about a swivel point by an optical trap, so that either the upper or lower channel is sealed and the flow is directed towards the open channel [12]. Another useful technique, based on a microfluidic system that uses optical tweezers, is known as a beam steerer. This works on the principle that an optically trapped microfluidic element, e.g. a silica microsphere, acts as a lens and refracts by varying degrees a signal beam as the particles move across it [13, 14]. Since optofluidics is still an emerging field, refining techniques to improve lab-on-a-chip devices continues; it is expected that optical manipulation effects will play an increasingly significant role.

References

[1] Grier D G 2003 A revolution in optical manipulation *Nature* **424** 810–6
[2] Curtis J E, Koss B A and Grier D G 2002 Dynamic holographic optical tweezers *Opt. Commun.* **207** 169–75
[3] Erickson D, Serey X, Chen Y-F and Mandal S 2011 Nanomanipulation using near field photonics *Lab Chip* **11** 995–1009
[4] Padgett M and Di Leonardo R 2011 Holographic optical tweezers and their relevance to lab on chip devices *Lab Chip* **11** 1196–205
[5] Mohanty S 2012 Optically-actuated translational and rotational motion at the microscale for microfluidic manipulation and characterization *Lab Chip* **12** 3624–36
[6] Ladavac K and Grier D G 2004 Microoptomechanical pumps assembled and driven by holographic optical vortex arrays *Opt. Express* **12** 1144–9
[7] Neale S L, MacDonald M P, Dholakia K and Krauss T F 2005 All-optical control of microfluidic components using form birefringence *Nat. Mater.* **4** 530–3
[8] Metzger N K, Mazilu M, Kelemen L, Ormos P and Dholakia K 2011 Observation and simulation of an optically driven micromotor *J. Opt.* **13** 044018
[9] Galajda P and Ormos P 2001 Complex micromachines produced and driven by light *Appl. Phys. Lett.* **78** 249–51
[10] Friese M E J, Nieminen T A, Heckenberg N R and Rubinsztein-Dunlop H 1998 Optical alignment and spinning of laser-trapped microscopic particles *Nature* **394** 348–50
[11] Terray A, Oakey J and Marr D W M 2002 Microfluidic control using colloidal devices *Science* **296** 1841–4
[12] Terray A, Oakey J and Marr D W M 2002 Fabrication of linear colloidal structures for microfluidic applications *Appl. Phys. Lett.* **81** 1555–7

[13] Domachuk P, Cronin-Golomb M, Eggleton B J, Mutzenich S, Rosengarten G and Mitchell A 2005 Application of optical trapping to beam manipulation in optofluidics *Opt. Express* **13** 7265–75

[14] Domachuk P, Omenetto F G, Eggleton B J and Cronin-Golomb M 2007 Optofluidic sensing and actuation with optical tweezers *J. Opt. A: Pure Appl. Opt.* **9** S129–133

Optical Nanomanipulation

David L Andrews and David S Bradshaw

Chapter 11

Vortex plasmons and light-induced ring currents

Previously, in chapter 8, we established that particles may circulate within a ring of a twisted beam. Recent theory on atoms or ions that undergo such motion is now examined for two cases. Both involve counter-propagating Laguerre-Gaussian beams; one in connection with surface plasmon optical vortices, and the other with light-induced ring currents.

Surface plasmon optical vortex

One of the most active areas of current research and development in the science of optics and optical materials is the burgeoning field of plasmonics. Most of this research concerns the light-induced excitation of *surface plasmons*: quantised charge oscillations that can form on the surface of a metal or at a metal interface with a dielectric solid. The strongly enhanced electric fields that such plasmons produce at boundaries and edges leads into a rich variety of applications [1–9]. On smooth metals, surface plasmons share quantum propagation features of both electrons and photons, and their character is strongly influenced by the specific electronic properties of the metal at optical frequencies. Indeed, this is reflected in another common designation, 'surface plasmon polaritons'—see, for example, chapter 12 of reference [10]. Under normal circumstances, the surface plasmon of a smooth metal cannot be excited directly with light from a vacuum, because of a sizeable mismatch between the momentum conveyed by a free-space photon and the momentum of a plasmon with the corresponding frequency. However, this problem can be overcome by arranging for light to impinge upon a metallic film by internal reflection inside a prism, wherein the photon momentum is larger than its vacuum value.

It transpires that localised non-propagating surface plasmons can be formed by exploiting the properties of counter-propagating optical vortex light, since such beams can instigate plasmonic excitation [11]. A suitable set-up is shown in figure 11.1, based on a Kretschmann prism configuration, where counter-propagating Laguerre-Gaussian beams undergo total internal reflection at a common position on

doi:10.1088/978-1-6817-4465-0ch11

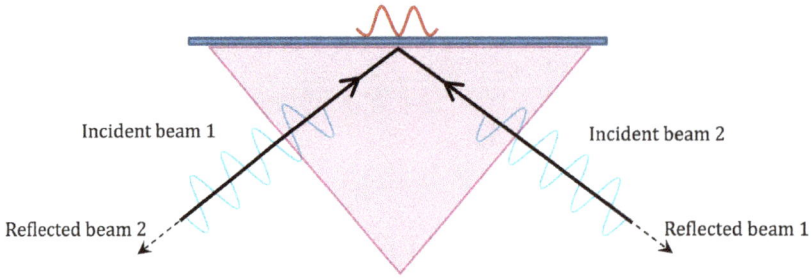

Figure 11.1. Plasmonic scheme based on the Kretschmann prism (pink triangle) configuration. Total internal reflection of two counter-propagating Laguerre-Gaussian beams (black arrows), at a planar dielectric interface with a metallic film (blue rectangle), creates a surface plasmon optical vortex (red wavy line).

(a)   (b)

Figure 11.2. (a) Typical trajectory (grey line) of an atom within a surface plasmon optical vortex. (b) Schematic of a light-induced ring current, generated by an ion circulating (red line) within 1D twisted optical molasses. Both schemes are based on two red-detuned, counter-propagating Laguerre-Gaussian beams: one with $l = 1$ and the other $l = -1$, and $p = 0$ in both cases. The dot represents the initial position of the atom or ion.

a prism face—this results in the formation of a surface plasmon optical vortex [12–14]. Whereas surface plasmons generally have a propagating character associated with linear momentum along the surface, the two-beam configuration provides for a cancellation of equal and opposite surface components so that the excitation is sustained at a fixed (localised) position, where the input beams meet. Conversely, however, when those beams have orbital angular momenta of equal and opposite sign, their positive addition provides a combined electrodynamic torque that drives the plasmon excitation, resulting in vortex behaviour. The transverse field distribution and angular momentum of the resulting plasmon can be visualised with near-field scanning optical microscopy [15]. Following formation of the surface plasmon optical vortex, any atom within the vicinity of the metallic surface may be trapped in the high intensity part of the optical vortex, due to the gradient force, and will undergo motion in a plane parallel to the surface as a result of the radiation force—a typical atom trajectory is shown in figure 11.2(a). It is important to note that any theory relating to this type of trapping belongs to the strong coupling regime, where light–matter interactions are large. Therefore, the perturbation methods introduced in chapter 2 cannot be applied to this plasmonic system.

1D twisted optical molasses

An ion trapped within a single Laguerre-Gaussian beam will undergo a helical trajectory due to the radiation force [16]: the usual linear push is combined with a circular motion that is either clockwise or anticlockwise depending on the positive or negative topological charge of the beam, respectively. In a scheme of counter-propagating Laguerre-Gaussian beams, known as 1D twisted optical molasses [17][1], the radiation forces responsible for pushing the ion downstream (i.e. in the propagation direction of the beam) will cancel out. As a result the motion of the ion is highly restricted in the axial direction. However, a light-induced ring current will occur (typically in the picoampere range, depending on the ion mass and charge), as shown by figure 11.2(b), since the ion still circulates in the plane of the other two dimensions.

References

[1] Zayats A V and Smolyaninov I I 2003 Near-field photonics: surface plasmon polaritons and localized surface plasmons *J. Opt. A: Pure Appl. Opt.* **5** S16–50

[2] Ozbay E 2006 Plasmonics: Merging photonics and electronics at nanoscale dimensions *Science* **311** 189–93

[3] Dragoman M and Dragoman D 2008 Plasmonics: Applications to nanoscale terahertz and optical devices *Prog. Quant. Electron* **32** 1–41

[4] Kawata S, Inouye Y and Verma P 2009 Plasmonics for near-field nano-imaging and superlensing *Nat. Photonics* **3** 388–94

[5] Atwater H A and Polman A 2010 Plasmonics for improved photovoltaic devices *Nat. Mater.* **9** 205–13

[6] Schuller J A, Barnard E S, Cai W, Jun Y C, White J S and Brongersma M L 2010 Plasmonics for extreme light concentration and manipulation *Nat. Mater.* **9** 193–204

[7] Kim J 2012 Joining plasmonics with microfluidics: from convenience to inevitability *Lab Chip* **12** 3611–23

[8] Zheng Y B, Kiraly B, Weiss P S and Huang T J 2012 Molecular plasmonics for biology and nanomedicine *Nanomedicine* **7** 751–70

[9] Sonntag M D, Klingsporn J M, Zrimsek A B, Sharma B, Ruvuna L K and Van Duyne R P 2014 Molecular plasmonics for nanoscale spectroscopy *Chem. Soc. Rev.* **43** 1230–47

[10] Novotny L and Hecht B 2006 *Principles of Nano-Optics* (Cambridge: Cambridge University Press)

[11] Tan P S, Yuan X-C, Lin J, Wang Q, Mei T, Burge R E and Mu G G 2008 Surface plasmon polaritons generated by optical vortex beams *Appl. Phys. Lett.* **92** 111108

[12] Lembessis V E, Babiker M and Andrews D L 2009 Surface optical vortices *Phys. Rev.* A **79** 011806

---

[1] In contrast, as introduced in chapter 4, counter-propagating Gaussian beams are typically used to form optical molasses.

[13] Andrews D L, Babiker M, Lembessis V E and Al-Awfi S 2010 Surface plasmons with phase singularities and their effects on matter *Phys. Status Solidi Rapid Res. Lett.* **4** 241–3

[14] Lembessis V E, Al-Awfi S, Babiker M and Andrews D L 2011 Surface plasmon optical vortices and their influence on atoms *J. Opt.* **13** 064002

[15] Shen Z, Hu Z J, Yuan G H, Min C J, Fang H and Yuan X C 2012 Visualizing orbital angular momentum of plasmonic vortices *Opt. Lett.* **37** 4627–9

[16] Carter A R, Babiker M, Al-Amri M and Andrews D L 2005 Transient optical angular momentum effects in light-matter interactions *Phys. Rev.* A **72** 043407

[17] Carter A R and Babiker M 2008 Twisted optical molasses for all-optical atomic cooling and trapping *Phys. Rev.* A **77** 043401

IOP Concise Physics

Optical Nanomanipulation

David L Andrews and David S Bradshaw

Chapter 12

Optical binding

Thus far, we have focused attention on the response of individual particles to optical forces. Provided each particle does not significantly occlude the light, then other particles within the same beam will respond in a similar manner. However, it emerges that additional, inter-particle forces may come into play (figure 12.1)—especially over distances where so-called dispersion forces apply. To understand these additional forces, the origin of the latter needs to be recognised.

The dispersion force: a comparison

Dispersion forces—also known as London forces [1–3], and principally repre-sented by the Casimir–Polder interaction—are the forces that exist between any electrically neutral and essentially non-polar particles, beyond the 'contact' region where their individual electronic wavefunctions would significantly overlap. Typically characterised by a monotonic inverse-sixth power dependence on separation, such inter-particle forces of attraction operate over a surprisingly large range of scale—their importance is not limited to atoms or molecules, but even extend to the Hamaker interactions [4] that hold together ceramic micro-particles. Indeed, these dispersion forces are responsible for the sustained integrity of most condensed phase matter; in their absence, repulsion between the outermost regions of electron density in separate particles would lead to instant material vaporisation. As Casimir and Polder showed [5], the distance dependence does eventually tail off into an inverse-seventh power, whose experimental proof verified that all such forces are ultimately produced by vacuum fluctuations. Associated with the latter, the fleeting presence of 'virtual' photons between the particles are responsible for the dispersion forces [6].

Returning to our main topic—and in clear distinction to the dispersion forces that occur whether or not any light is present[1]—there are, in fact, additional forces that

---

[1] It is generally understood that light is comprised of 'real' photons.

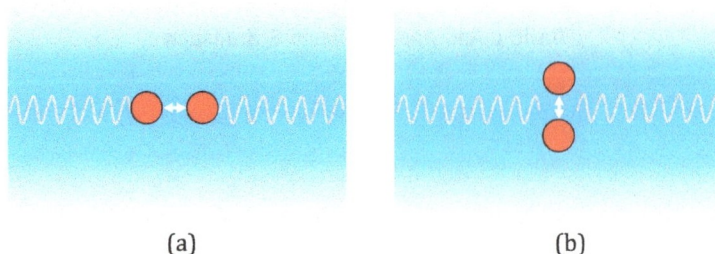

(a)   (b)

Figure 12.1. Diagrams depicting (a) longitudinal and (b) transverse optical binding. White double-headed arrows represent inter-particle interactions and the grey wavy lines denote the throughput of a Gaussian beam.

arise between particles, specifically connected with the presence of light. In many experiments these optically induced inter-particle forces prove to be much larger than the dispersion forces, and although they both diminish with distance, the former can still be significant at separations that are similar in magnitude to the optical wavelength. The effect is now commonly known by the term 'optical binding', coined by the authors of some of the earliest experimental work [7]. Nonetheless, as we shall see, the form of this interaction does not always have the quality of attraction that 'binding' usually signifies.

Theory of optical binding

Classical description. Before describing the quantum mechanism, it is helpful to reason through the origin of optical binding forces in essentially classical terms. Consider a pair of particles, in reasonably close proximity, that are irradiated by light of sufficient intensity. Each particle (which intrinsically has the property of polarisability—meaning that its constituent charges move in response to the electromagnetic radiation) will exhibit an oscillating dipole, due to the presence of the laser light. These oscillations will, however, generally differ in phase due to the difference in positioning of the two particles with respect to the light waves. The oscillating dipoles create secondary fields which are experienced by the partner particle. The result is mutual interaction between the particles which, depending on the inter-particle disposition, may be either in-phase or out-of-phase—leading to either repulsive or attractive forces, respectively. So we can anticipate that 'optical binding' forces will fall off with distance, cycling between progressively diminishing positive and negative values as the separation increases. This scheme of rationalisation is illustrated in figure 12.2.

Quantum description. A rigorous theory of optical binding calls for the deployment of quantum electrodynamical methods, wherein all inter-particle forms of interaction are mediated by 'virtual' photon exchange. First formulated by Thirunamachandran for specific application to molecules [8], this theory provides a definitive mechanism and a quantitative basis for characterising the effect, as qualitatively described above. The essential mechanism is pairwise forward Rayleigh scattering [9] (analogous to the gradient force of single-particle optical trapping),

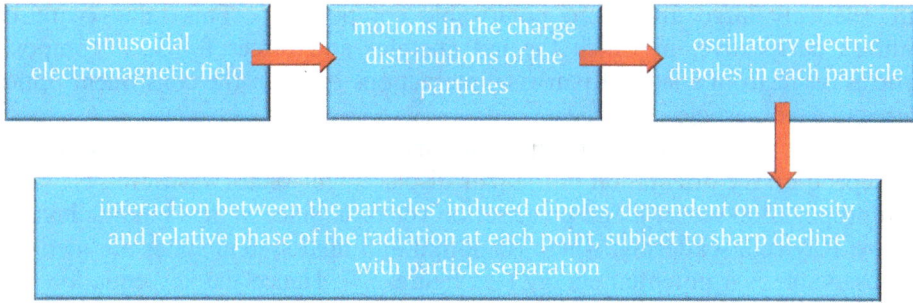

Figure 12.2. Outline of the classical description of optical binding.

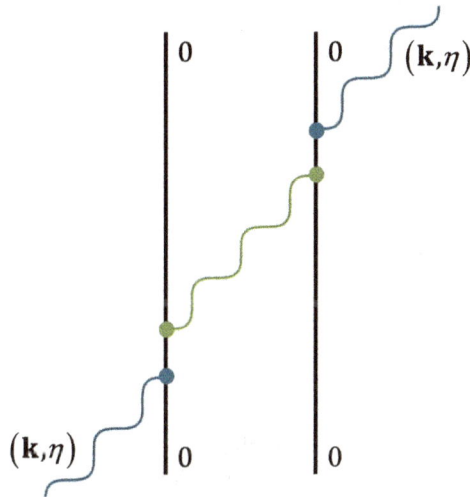

Figure 12.3. Feynman diagram illustrating optical binding. Blue and green wavy lines respectively represent 'real' and 'virtual' photons, and the dots denote light–matter interaction. The annihilated and created throughput beam ('real') photons are identical, and both particles begin and end in the ground state 0.

which involves the annihilation of one laser beam photon by one particle and re-creation of an identical photon, repopulating the beam, by the neighbour. The two events are causally linked by 'virtual' photon coupling, as shown by the Feynman diagram of figure 12.3. By summing the matrix elements corresponding to each such diagram, a pair interaction energy can be evaluated, whose relative position-dependence allows the evaluation of local forces. Remarkably, this theory proves to be applicable to much more than molecules, including dielectric particles of nanometre and micron size, and even extending to metal nanoparticles. It emerges that this responsive force depends on a product of the polarisabilities for the interacting pair, and as such is much more amenable to experimental observation with relatively large and more polarisable components[2]. This tallies with the quantum perspective, in which the net energies and forces for such composite particles result from a sum of pairwise interactions between the constituent optical centres in each particle.

In addition to a rapidly declining, oscillatory dependence on distance, local forces also depend on the orientation of the pair displacement vector **s** with respect to the beam propagation vector **k** and polarisation vector **e**. Since optical binding involves forward Rayleigh scattering, where the initial and final system states are identical, the observable is a potential energy (see figure 2.3). Hence the key equations, for 'longitudinal' (where **s**||**k**, **s**⊥**e**) and 'transverse' (**s**⊥**k**, **s**||**e**) optical binding respectively—as depicted in figure 12.1—are given as follows;

$$\Delta E = \left(\frac{I\alpha_A\alpha_B}{4\pi\varepsilon_0^2 cs^3}\right)\left((1-k^2s^2)\cos(2ks) + ks\,\sin(2ks)\right), \tag{12.1}$$

$$\Delta E = -\left(\frac{I\alpha_A\alpha_B}{2\pi\varepsilon_0^2 cs^3}\right)\left(\cos(ks) + ks\,\sin(ks)\right). \tag{12.2}$$

Here, α_A and α_B are respectively the scalar polarisabilities of particle A and B, and $s = |\mathbf{s}|$ is the separation distance between the two particles. The optical binding force is then determined by inserting equations (12.1) and (12.2) into $\mathbf{F} = -\partial(\Delta E)/\partial s$; the intricate result (found elsewhere in a generalised form [9]) will show that optical binding has a short-range inverse-fourth power distance dependence, i.e. s^{-4}. In this chapter the throughput beam is assumed Gaussian; however, optical binding using structured light has also been examined [10–15].

How the potential energy varies with particle separation is illustrated in figure 12.4 for the two configurations; this clearly shows that the deeper potential wells, and hence stronger forces, are associated with particles positioned along the beam axis. The most striking aspect, in each case, is a series of separation distances at which the energy is a minimum, such that the optically induced force is neither attractive nor repulsive, but vanishes (since the gradient of the potential energy curve is zero). At each of these points, two particles can be held in a sustainable equilibrium position, well beyond material contact. In the case of the closest, deepest potential energy minimum the arrangement can be stable against small perturbations; indeed, by disturbing the system it is possible to observe spring-like mechanical oscillations as the optically bound equilibrium is restored after displacement. Treating such vibrations as having a simple harmonic form enables an effective force constant to be calculated for the bound pair [16].

Potential energy landscapes

Most of the theory on optical binding concerns laser-induced interactions between spherical particles. For progressively more intricate systems, the

---

[2] In the Drude model, the magnitude of a scalar polarisability scales with particle volume.

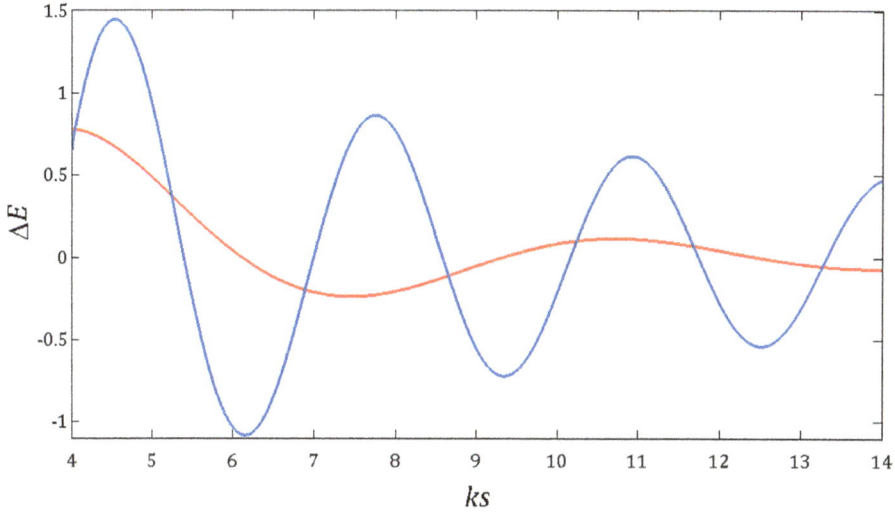

Figure 12.4. Graph of potential energy, ΔE, (with arbitrary scale) against pair separation measured in terms of ks, for longitudinal (blue line) and transverse (red line) optical binding; k is given a typical value of 3×10^{-7} m^{-1}. Notice that both oscillating curves are damped as the separation distance increases.

complexity of the theory quickly escalates, compounded by additional difficulty when three or more particles experience collective optical binding. For example, even a pair of cylindrical particles has seven degrees of spatial freedom; the associated analysis, based on carbon nanotubes, was first tackled by the present authors [17][3], Analogous to trapping, the optical binding of cylindrical nano-particles is experimentally more readily achievable because one of its dimensions extends into the micron size range. Nevertheless, for the common case of spherical particles, a considerable amount of groundwork has been accomplished—leading to the mapping of potential energy landscapes [18], such as those illustrated in figure 12.5. Within such energy surfaces the particles will reside in the minima, at positions where the optical binding force is null, so that certain stable arrangements can be maintained with fixed particle–particle separations.

This capacity for particles to be held together by light, in stable and non-contact geometries, bears a superficial similarity to the way in which atoms are held together by Coulombic forces, in molecules of distinct geometry. Recognition of this analogy has led to some, perhaps over-enthusiastic, proclamations of a new form of 'optical matter' [19]. However, this exaggeration does not detract from genuine possibilities for optical self-assembly, driven by the forces of optical binding. In one experiment, the formation of two- and three-dimensional clusters has been observed with

---

[3] For nanotubes of 200 nm length and 0.4 nm radius, separated by a distance of 2 nm and irradiated by a light beam with 1×10^{12} W cm^{-2} intensity, it was determined that the optical binding force ranges between 10^{-5} and 10^{-12} N (according to geometry).

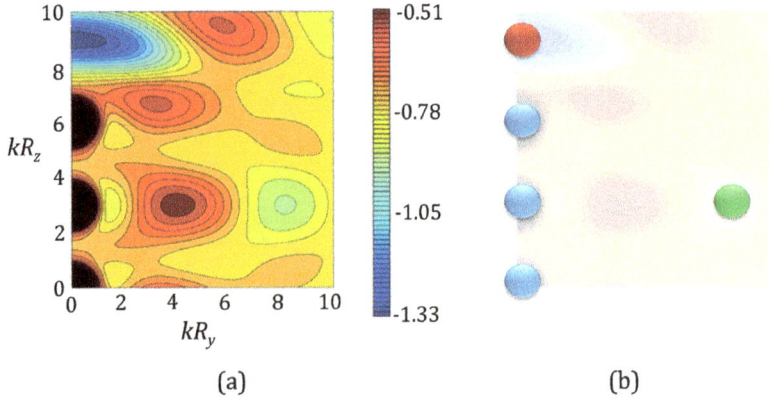

Figure 12.5. (a) Potential energy landscape for the pairwise optical binding of four identical particles in the y, z-plane. Polarisation and wave-vector of the input beam is along the x- and y-axis, respectively. Three particles are already set in positions fixed by optical binding potentials, as represented by the black circular shapes at $[kR_y, kR_z] = [0, 0]$, $[0, 3]$, $[0, 6]$. The fourth particle is variably positioned and depends on the new potential energy surface; as a result it most likely resides in a minimum at $[0, 9]$ or $[8, 3]$. Colour scale (arbitrary units) on the right. (b) Two stable cluster arrangements based on the potential energy landscape (a). The three fixed particles are represented by blue circles. Red circle signifies the fourth particle in its most favourable position (within the absolute minimum), while green circle indicates the fourth particle in its other stable location (a local minimum). Such clusters have been observed experimentally, as shown in reference [20]. Image (a) is adapted from reference [18]. Reprinted with permission. Copyright 2008 by the American Physical Society.

arrangements predicted by optical landscape theory [20]. In another, long chains of sub-micron polystyrene spheres have been created with hexagonal symmetry [21]. Formation of these lattices is a consequence of the ensemble adopting a minimum of net potential energy; the particles are naturally separated into non-contact positions that are distinct from the close-packing exhibited in many familiar physical structures, such as the optical lattice structures discussed in chapter 7 (where inter-particle interactions are minimal).

In practice, the optical binding of nanomaterials is at present generally restricted to metal nanoparticles [20, 22–25]—since their larger polarisabilities, compared to dielectric substances, enable the production of a binding force that is more amenable to experimental measure. Moreover, as with optical tweezing, most experiments involve particles immersed in a liquid medium [25–28] (although, in connection with microdroplets, optical binding in air has also been reported [29, 30]). In general, any beam or beams that are involved in producing optical binding effects will usually have first acted as traps for the particles individually. It is therefore necessary to also account for the fact that, alongside the gradient force which moves the particles within useful range of each other, there is also a radiation force tending to push each of them along the beam axis. Such potential problems with radiation forces, in optical binding experiments, can be solved by using counter-propagating beams [31, 32]; in a set-up remarkably similar to the first optical trap created by Ashkin in 1970 [33].

References

[1] Eisenschitz R and London F 1930 Über das Verhältnis der van der Waalsschen Kräfte zu den homöopolaren Bindungskräften *Z. Physik* **60** 491–527

[2] London F 1930 Zur Theorie und Systematik der Molekularkräfte *Z. Physik* **63** 245–79

[3] London F 1937 The general theory of molecular forces *J. Chem. Soc. Faraday Trans.* **33** 8–26

[4] Hamaker H C 1937 The London-van der Waals attraction between spherical particles *Physica* **4** 1058–72

[5] Casimir H B G and Polder D 1948 The influence of retardation on the London–van der Waals forces *Phys. Rev.* **73** 360–72

[6] Andrews D L and Bradshaw D S 2014 The role of virtual photons in nanoscale photonics *Ann. Phys. (Berlin)* **526** 173–86

[7] Burns M M, Fournier J-M and Golovchenko J A 1989 Optical binding *Phys. Rev. Lett.* **63** 1233–6

[8] Thirunamachandran T 1980 Intermolecular interactions in the presence of an intense radiation field *Mol. Phys.* **40** 393–9

[9] Bradshaw D S and Andrews D L 2005 Optically induced forces and torques: Interactions between nanoparticles in a laser beam *Phys. Rev.* A **72** 033816

[10] Bradshaw D S and Andrews D L 2005 Interactions between spherical nanoparticles optically trapped in Laguerre-Gaussian modes *Opt. Lett.* **30** 3039–41

[11] Karásek V and Zemánek P 2007 Analytical description of longitudinal optical binding of two spherical nanoparticles *J. Opt. A: Pure Appl. Opt.* **9** S215–S220

[12] Karásek V, Čižmár T, Brzobohatý O, Zemánek P, Garcés-Chávez V and Dholakia K 2008 Long-range one-dimensional longitudinal optical binding *Phys. Rev. Lett.* **101** 143601

[13] Rodríguez J and Andrews D L 2008 Optical binding and the influence of beam structure *Opt. Lett.* **33** 2464–6

[14] Karásek V, Brzobohatý O and Zemánek P 2009 Longitudinal optical binding of several spherical particles studied by the coupled dipole method *J. Opt. A: Pure Appl. Opt.* **11** 034009

[15] Rodríguez J and Andrews D L 2009 Influence of the state of light on the optically induced interparticle interaction *Phys. Rev.* A **79** 022106

[16] Metzger N, Marchington R, Mazilu M, Smith R, Dholakia K and Wright E 2007 Measurement of the restoring forces acting on two optically bound particles from normal mode correlations *Phys. Rev. Lett.* **98** 068102

[17] Andrews D L and Bradshaw D S 2005 Laser-induced forces between carbon nanotubes *Opt. Lett.* **30** 783–5

[18] Rodríguez J, Dávila Romero L C and Andrews D L 2008 Optical binding in nanoparticle assembly: Potential energy landscapes *Phys. Rev.* A **78** 043805

[19] Burns M M, Fournier J-M and Golovchenko J A 1990 Optical matter: crystallization and binding in intense optical fields *Science* **249** 749–54

[20] Yan Z, Gray S K and Scherer N F 2014 Potential energy surfaces and reaction pathways for light-mediated self-organization of metal nanoparticle clusters *Nat. Commun.* **5** 3751

[21] Mellor C D, Fennerty T A and Bain C D 2006 Polarization effects in optically bound particle arrays *Opt. Express* **14** 10079–88

[22] Demergis V and Florin E-L 2012 Ultrastrong optical binding of metallic nanoparticles *Nano Lett.* **12** 5756–60

[23] Yan Z, Shah R A, Chado G, Gray S K, Pelton M and Scherer N F 2013 Guiding spatial arrangements of silver nanoparticles by optical binding interactions in shaped light fields *ACS Nano* **7** 1790–802

[24] Yan Z, Manna U, Qin W, Camire A, Guyot-Sionnest P and Scherer N F 2013 Hierarchical photonic synthesis of hybrid nanoparticle assemblies *J. Phys. Chem. Lett.* **4** 2630–6

[25] Bowman R W and Padgett M J 2013 Optical trapping and binding *Rep. Prog. Phys.* **76** 026401

[26] Mohanty S K, Andrews J T and Gupta P K 2004 Optical binding between dielectric particles *Opt. Express* **12** 2746–53

[27] Dholakia K and Zemanek P 2010 Gripped by light: Optical binding *Rev. Mod. Phys.* **82** 1767–91

[28] Brzobohatý O, Čižmár T, Karásek V, Šiler M, Dholakia K and Zemánek P 2010 Experimental and theoretical determination of optical binding forces *Opt. Express* **18** 25389–402

[29] Guillon M, Moine O and Stout B 2006 Longitudinal optical binding of high optical contrast microdroplets in air *Phys. Rev. Lett.* **96** 143902

[30] Guillon M and Stout B 2008 Optical trapping and binding in air: Imaging and spectroscopic analysis *Phys. Rev.* A **77** 023806

[31] Tatarkova S A, Carruthers A E and Dholakia K 2002 One-dimensional optically bound arrays of microscopic particles *Phys. Rev. Lett.* **89** 283901

[32] Singer W, Frick M, Bernet S and Ritsch-Marte M 2003 Self-organized array of regularly spaced microbeads in a fiber-optical trap *J. Opt. Soc. Am.* B **20** 1568–74

[33] Ashkin A 1970 Acceleration and trapping of particles by radiation pressure *Phys. Rev. Lett.* **24** 156–9

Chapter 13

Past, present and future

Since ancient times, humankind has wrestled with conceiving the true nature of light and its interactions. Some of the earliest attempts to understand vision were based on light supposedly being involved in a tangible influence, in ways that would be hard to square with any modern scientific view. A classic example from first century B.C. is to be found in the all-embracing explanation of the world entitled *De Rerum Natura* by Lucetius [1]. Such views owed their origin to attempted rationalisations of how distant objects could possibly engage with the eye, without mechanical contact. Once such ideas were dispelled, through renaissance advances in science, it was found to be more than a curiosity when the notion of mechanical interactions was eventually put on a firm scientific basis, primarily through the pioneering work on theory by Bartoli, Maxwell, Poynting and others. It was no surprise, however, to recognize that on Earth the levels of force associated with sunlight—or indeed any other common source of light, were well below the threshold of everyday perception. As we observed in chapter 1, the first experiments confirming that theoretical work came in 1901, in work by Lebedev [2] as well as by Nichols and Hull [3]. However, it was with the arrival of the laser more than half a century later that practical applications in the form of optical trapping started to emerge.

Fifty years further on, the transformed field of optical forces has been enormously expanded, and is even now being extended and reinvigorated in a variety of wholly unanticipated directions. A number of these reflect advances in nanotechnology, which now enable the manipulation of the tiniest amounts of matter with unprecedented precision. But it is not only these developments that have unfolded new applications; advances in laser technology have enabled the circumvention of long-held tenets in conventional optics—such as the impossibility of beating the diffraction limit [4], the capacity to produce beams that are diffraction-free [5] or appear to defy the law of rectilinear propagation [6]—even the possibility of beams with polarisation states that vary in free space [7]. As the range of different mechanisms at work in optical manipulation continues to evolve, our aim in this book has been to make clear the principles and formulations of theory. It is our hope that by providing a systematic description of past and current innovations, the same framework will also provide a basis to understand new effects that undoubtedly lie, at present, just beyond the horizon.

References

[1] Carus T L and Stallings A 2007 *The Nature of Things* (London: Penguin Books)
[2] Lebedev P N 1901 Experimental examination of light pressure *Ann. Phys. (Berlin)* **6** 433
[3] Nichols E F and Hull G F 1901 A preliminary communication on the pressure of heat and light radiation *Phys. Rev.* **13** 307–20
[4] Hell S W and Wichmann J 1994 Breaking the diffraction resolution limit by stimulated emission: stimulated-emission-depletion fluorescence microscopy *Opt. Lett.* **19** 780–2
[5] Bouchal Z and Olivík M 1995 Non-diffractive vector bessel beams *J. Mod. Opt.* **42** 1555–66
[6] Siviloglou G A, Broky J, Dogariu A and Christodoulides D N 2007 Observation of accelerating airy beams *Phys. Rev. Lett.* **99** 213901
[7] Schulz S A, Machula T, Karimi E and Boyd R W 2013 Integrated multi vector vortex beam generator *Opt. Express* **21** 16130–41

Chapter 14

Bibliography

Chapters 1 & 2

Andrews D L 2015 *Photonics, Fundamentals of Photonics and Physics* (New York: Wiley)

Craig D P and Thirunamachandran T 1998 *Molecular Quantum Electrodynamics* (Mineola, NY: Dover)

Demtröder W 2010 *Atoms, Molecules and Photons: An Introduction to Atomic-, Molecular- and Quantum-Physics* (Berlin: Springer)

Grynberg G, Aspect A and Fabre C 2010 *Introduction to Quantum Optics: From the Semi-Classical Approach to Quantized Light* (Cambridge: Cambridge University Press)

Saleem M 2016 *Quantum Mechanics* (Bristol: Institute of Physics Publishing)

Chapter 3

Jones P H, Maragò O M and Volpe G 2015 *Optical Tweezers: Principles and Applications* (Cambridge: Cambridge University Press)

Padgett M J, Molloy J and McGloin D 2010 *Optical Tweezers: Methods and Applications* (Boca Raton, FL: CRC Press)

Chapter 4

Metcalf H J and van der Straten P 2012 *Laser Cooling and Trapping* (New York: Springer)

Pethick C J and Smith H 2008 *Bose–Einstein Condensation in Dilute Gases* 2nd edn (Cambridge: Cambridge University Press)

Ueda M 2010 *Fundamentals and New Frontiers of Bose–Einstein Condensation* (Singapore: World Scientific)

Chapters 5 & 6

Ashkin A 2006 *Optical Trapping and Manipulation of Neutral Particles using Lasers* (Singapore: World Scientific)

Gouesbet G and Gréhan G 2011 *Generalized Lorenz–Mie Theories* (Berlin: Springer)

Jones P H, Maragò O M and Volpe G 2015 *Optical Tweezers: Principles and Applications* (Cambridge: Cambridge University Press)

Padgett M J, Molloy J and McGloin D 2010 *Optical Tweezers: Methods and Applications* (Boca Raton, FL: CRC Press)

Chapter 7

Glückstad J and Palima D 2009 *Generalized Phase Contrast: Applications in Optics and Photonics* (Berlin: Springer)

Lewenstein M, Sanpera A and Ahufinger V 2012 *Ultracold Atoms in Optical Lattices: Simulating Quantum Many-Body Systems* (Oxford: Oxford University Press)

Chapters 8 & 9

Allen L, Barnett S M and Padgett M J 2003 *Optical Angular Momentum* (Boca Raton, FL: CRC Press)

Andrews D L 2011 *Structured Light and its Applications: An Introduction to Phase-Structured Beams and Nanoscale Optical Forces* (New York: Academic)

Andrews D L and Babiker M 2012 *The Angular Momentum of Light* (Cambridge: Cambridge University Press)

Gbur G J 2016 *Singular Optics* (Boca Raton, FL: CRC Press)

Chapter 10

Fainman Y, Lee L, Psaltis D and Yang C 2010 *Optofluidics: Fundamentals, Devices, and Applications* (New York: McGraw-Hill)

Monat C, Domachuk P and Eggleton B J 2007 Integrated optofluidics: a new river of light *Nat. Photonics* **1** 106–14

Chapter 11

Kawata S 2001 *Near-Field Optics and Surface Plasmon Polaritons* (Berlin, Heidelberg: Springer)

Maier S A 2007 *Plasmonics: Fundamentals and Applications* (New York: Springer)

Novotny L and Hecht B 2006 *Principles of Nano-Optics* (Cambridge: Cambridge University Press)

Sarid D and Challener W A 2010 *Modern Introduction to Surface Plasmons: Theory, Mathematica Modeling, and Applications* (Cambridge: Cambridge University Press)

Chapter 12

Salam A 2010 *Molecular Quantum Electrodynamics. Long-Range Intermolecular Interactions* (Hoboken, NJ: Wiley)

Taylor J M 2011 *Optical Binding Phenomena: Observations and Mechanisms* (Berlin, Heidelberg: Springer)

www.ingramcontent.com/pod-product-compliance
Lightning Source LLC
Chambersburg PA
CBHW081550220326
41598CB00036B/6630